Property Rights and Governance in Artisanal and Small-Scale Mining

Disputes and dispossession of property rights in the mining sector are causes of injustice, violence, and forced resettlement around the world. This comprehensive volume examines mining, particularly what is often called 'Artisanal and Small-Scale Mining', from a perspective of governance and rights. It focuses on rights to land, natural resources, and other forms of material 'property'. Many projects, policies, and laws targeting artisanal and small-scale mining are embedded in problematic conceptual and institutional frameworks that implicitly stigmatise and discipline artisanal and small-scale miners. This collection takes a critical look at notions of property to destabilise some of these frameworks.

The chapters in this book are notable for their recognition of the agency of artisanal miners and 'local communities' within the uneven hierarchies in which they are embedded, and their acknowledgement of the difficulties of state regulation of such a complex set of issues. The authors use a variety of theoretical tools, engaging with political economy, political ecology, classical economic theory, and socio-cultural concepts derived from ethnographic methods.

This book includes insightful case studies from Colombia, Democratic Republic of Congo, Ghana, Mongolia, South Africa, and Zambia, and is an important resource for academics, development practitioners, and policy-makers. It was originally published online as a special issue of *Third World Thematics*.

Chris Huggins is Assistant Professor in the School of International Development and Global Studies at the University of Ottawa, Canada. His research focuses on the political economy of natural resource management in Africa. He is author of *Agricultural Reform in Rwanda: Authoritarianism, Markets and Zones of Governance* (2017).

ThirdWorlds

Edited by Shahid Qadir, *University of London, UK*

ThirdWorlds will focus on the political economy, development and cultures of those parts of the world that have experienced the most political, social, and economic upheaval, and which have faced the greatest challenges of the postcolonial world under globalisation: poverty, displacement and diaspora, environmental degradation, human and civil rights abuses, war, hunger, and disease.

ThirdWorlds serves as a signifier of oppositional emerging economies and cultures ranging from Africa, Asia, Latin America, Middle East, and even those 'Souths' within a larger perceived North, such as the U.S. South and Mediterranean Europe. The study of these otherwise disparate and discontinuous areas, known collectively as the Global South, demonstrates that as globalisation pervades the planet, the south, as a synonym for subalterity, also transcends geographical and ideological frontier.

The most recent titles include:

Class Dynamics of Development
Edited by Jonathan Pattenden, Liam Campling, Satoshi Miyamura and Benjamin Selwyn

Third World Approaches to International Law
Edited by Amar Bhatia, Usha Natarajan, John Reynolds and Sujith Xavier

Fragility, Aid, and State-building
Understanding Diverse Trajectories
Edited by Rachel M. Gisselquist

Rural Transformations and Agro-Food Systems
The BRICS and Agrarian Change in the Global South
Edited by Ben M. McKay, Ruth Hall and Juan Liu

Sustainable Development in Africa-EU relations
Edited by Mark Langan and Sophia Price

New Mechanisms of Participation in Extractive Governance
Between Technologies of Governance and Resistance Work
Edited by Esben Leifsen, Maria-Therese Gustafsson, María Antonieta Guzmán-Gallegos and Almut Schilling-Vacaflor

The Development Dictionary @25
Post-Development and its Consequences
Edited by Aram Ziai

Power and Illicit Drugs in the Global South
Edited by Maziyar Ghiabi

Hybridity in Peacebuilding and Development
A critical and reflexive approach
Edited by Lia Kent, Miranda Forsyth, Sinclair Dinnen, Joanne Wallis and Srinjoy Bose

The Agency of the Governed in the Global South
Normative and Institutional Change
Edited by Anke Draude

Innovations in 'Sport for Development and Peace' Research
Edited by Megan Chawansky, Lyndsay M. C. Hayhurst, Mary G. McDonald and Cathy van Ingen

Global Debt Dynamics
Crises, Lessons, Governance
Edited by Andreas Antoniades and Ugo Panizza

Developmental States beyond East Asia
Edited by Jewellord T. Nem Singh and Jesse Salah Ovadia

Gendering Security and Insecurity
Post/Neocolonial Security Logics and Feminist Interventions
Edited by Navtej Purewal and Sophia Dingli

Property Rights and Governance in Artisanal and Small-Scale Mining
Critical Approaches
Edited by Chris Huggins

For more information about this series, please visit:
https://www.routledge.com/series/TWQ

Property Rights and Governance in Artisanal and Small-Scale Mining

Critical Approaches

Edited by

Chris Huggins

Routledge
Taylor & Francis Group
LONDON AND NEW YORK

First published 2020
by Routledge
2 Park Square, Milton Park, Abingdon, Oxon, OX14 4RN

and by Routledge
52 Vanderbilt Avenue, New York, NY 10017

Routledge is an imprint of the Taylor & Francis Group, an informa business

First issued in paperback 2021

British Library Cataloguing in Publication Data
A catalogue record for this book is available from the British Library

ISBN 13: 978-0-367-23673-1 (hbk)
ISBN 13: 978-1-03-209045-0 (pbk)

Typeset in Myriad Pro
by RefineCatch Limited, Bungay, Suffolk

Publisher's Note
The publisher accepts responsibility for any inconsistencies that may have arisen during the conversion of this book from journal articles to book chapters, namely the inclusion of journal terminology.

Contents

Citation Information

The chapters in this book were originally published in *Third World Thematics*, volume 1, issue 2 (April 2016). When citing this material, please use the original page numbering for each article, as follows:

Chapter 1

Artisanal and small-scale mining: critical approaches to property rights and governance
Chris Huggins
Third World Thematics, volume 1, issue 2 (April 2016), pp. 151–164

Chapter 2

Revisiting the interconnections between research strategies and policy proposals: reflections from the artisanal and small-scale mining sector in Africa
Bonnie Campbell
Third World Thematics, volume 1, issue 2 (April 2016), pp. 165–183

Chapter 3

The politics of artisanal and small-scale mining in Mongolia
Pascale Hatcher
Third World Thematics, volume 1, issue 2 (April 2016), pp. 184–201

Chapter 4

Property rights and large-scale mining: overlapping claims at and around mining sites at the Democratic Republic of Congo and Zambia
Sarah Katz-Lavigne
Third World Thematics, volume 1, issue 2 (April 2016), pp. 202–217

Chapter 5

'Custom' and fractured 'community': mining, property disputes and law on the platinum belt, South Africa
Sonwabile Mnwana
Third World Thematics, volume 1, issue 2 (April 2016), pp. 218–234

Chapter 6

Disputes over gold mining and dispossession of local afrodescendant communities from the Alto Cauca, Colombia
Irene Vélez-Torres
Third World Thematics, volume 1, issue 2 (April 2016), pp. 235–248

Chapter 7

Different faces of access control in a Congolese gold mine
Sara Geenen and Klara Claessens
Third World Thematics, volume 1, issue 2 (April 2016), pp. 249–266

Chapter 8

Artisanal gold mining in Kejetia (Tongo, Northern Ghana): a three-dimensional perspective
Esther van de Camp
Third World Thematics, volume 1, issue 2 (April 2016), pp. 267–283

Notes on Contributors

Esther van de Camp is a PhD candidate at the Institute of Cultural Anthropology and Development Sociology at Leiden University, the Netherlands. Her research is part of the Gold Matters project, carrying out and comparing ASGM-research in South America, West Africa, and East Africa. Her field research takes place in Uganda.

Bonnie Campbell is Associate Professor of Political Economy in the Department of Political Science at the University of Quebec in Montreal (UQAM), Canada. At UQAM, she is the Director of the Interdisciplinary Centre for Research in International Development and Society (CIRDIS), and also the Director of the Research Group on Mining Activities in Africa. Professor Campbell has been a Member of the Royal Society of Canada since 2012.

Klara Claessens is a Postdoctoral Researcher in the Division of Geography and Tourism at KU Leuven, Belgium. Her research focuses on land relations, natural resource management, agrarian change, and local politics in the Great Lakes Region in Central Africa. She is interested in land relations and changing patterns of land access and control in eastern Democratic Republic of Congo (DRC).

Sara Geenen is a Lecturer in International Development, Globalization and Poverty at the Institute of Development Policy and Management (IOB) at the University of Antwerp, Belgium. She has published on artisanal gold mining and company–community conflicts in South Kivu, DRC. Her current research interests lie in the global and local development dimensions of extractivist projects, addressing questions about more socially responsible and inclusive forms of globalization.

Pascale Hatcher is a Senior Lecturer at the University of Canterbury, New Zealand. Her research focuses on the political economy of mining in the developing world, focusing specifically on the political economy of large-scale mining in Mongolia (the site of what are expected to become two of the world's largest mines).

Chris Huggins is Assistant Professor in the School of International Development and Global Studies at the University of Ottawa, Canada. His research focuses on the political economy of natural resource management in Africa. He is the author of *Agricultural Reform in Rwanda: Authoritarianism, Markets and Zones of Governance* (2017).

Sarah Katz-Lavigne is a PhD candidate in the International Development Policy stream at the Norman Paterson School of International Affairs (NPSIA) at Carleton University, Canada. Her research centres on multinational corporation involvement in mineral extraction in sub-Saharan Africa, particularly the DRC, applying a property rights lens.

Sonwabile Mnwana is Associate Professor of Sociology and Head of Department at the University of Fort Hare, South Africa. He is also a Research Associate at the Society, Work and Development Institute (SWOP) at the University of the Witwatersrand, South Africa. Broadly, his research focuses on meanings of land, large scale natural resource extraction, and rural social change.

Irene Vélez-Torres is Associate Professor in EIDENAR at the Universidad del Valle, Colombia. She has worked on the critical analysis of socio-environmental conflicts, rural–urban migration, ethnicity, and social movements in Latin America. She has explored participatory and interdisciplinary methodologies in order to create knowledge that is valuable for communities as well as for academia.

Artisanal and small-scale mining: critical approaches to property rights and governance

Chris Huggins

ABSTRACT

Disputes and dispossession of property rights in the mining sector are causes of injustice, violence and forced displacement around the world. Many artisanal and small-scale mining policy initiatives are embedded in problematic conceptual and institutional frameworks that implicitly stigmatise and discipline artisanal and small-scale miners. This collection takes a critical look at notions of property to destabilise some of these frameworks. Contributions are notable for their recognition of the agency of artisanal miners and 'local communities' within the uneven hierarchies in which they are embedded, and their acknowledgement of the difficulties of state regulation of such a complex set of issues.

Introduction

This collection examines mining, particularly what is often called 'Artisanal and Small-Scale Mining' (ASM), from a perspective of governance and rights. It focuses especially on rights to land, natural resources and other forms of material 'property'. ASM is hugely important, in terms of employment and mineral production, in many countries. There are at least 20–30 million artisanal and small-scale miners globally, with more than eight million of them working in Africa.[1] In African states, ASM is often responsible for more mineral production than industrial mining; in Latin America, even in the leftist states, there has been a multiplication of mining sites and an associated increase in conflicts, often violent, between mining firms, artisanal miners, local citizens and other actors over access to mineralised land and environmental impacts; and new mineral industries have recently sprung up in Central Asia, for example, accompanied by ASM. There has accordingly, over the past 15 years, been a rapid growth in academic and policy-oriented research and documentation on ASM. Some of this has resulted in important debates on how miners may be 'supported' to work more effectively, safely and profitably. Filling a glaring gap in the literature, the role of women in ASM has recently been critically examined.[2] However, some of the ASM literature has been criticised for various biases including a pro-industrial mining stance, a statist orientation that tends to privilege taxation and formalisation above the material and social interests of individual miners; and a simplistic conceptual framework that assumes various challenges can be overcome through legislation, 'formalization' or moratoriums on mining in certain areas. More fundamentally, given the 'radical' challenge that artisanal miners pose to industrial mining and property rights regimes, much of the discourse around ASM is seen by critics as a form of discursive regulation or discipline, 'a response designed to contain, reframe and re-direct the anticolonial challenge posed by, or potentially posed by, these subjects'.[3]

This collection takes a critical and multidisciplinary approach. It has resulted from a two-session panel at the conference of the Netherlands Academy on Land Governance for Equitable and Sustainable Development (LANDac) at Utrecht University in 2015. The panel, situated within a conference focused largely on the phenomenon of large-scale commercial acquisition of land in the Global South, or the 'foreignization of space',[4] highlighted the diversity of perspectives and lack of common action on the crisis within mineral-rich zones. With the exception of policies on free prior informed consent (FPIC) which only affect areas with 'indigenous' populations (as discussed in this collection), the mining sector has seen not the kinds of a high-profile multi-stakeholder push for *global* 'voluntary guidelines' as has been seen in the land sector.[5] Instead, a more complex, ad hoc, and regionally specific set of responses to particular challenges (such as 'conflict minerals') may be seen.

The ASM sector, and academic and policy responses to the challenges it represents, has numerous conceptual connections to broader themes of natural resources management, and international development more generally. At a general level, some of the key policy questions in the minerals sector revolve around the roles and responsibilities of the state, members of local communities, international NGOs, and the private sector. The first question is whether particular resources should be extracted: do the benefits to local livelihoods, or national public coffers, outweigh the negative environmental, social and health impacts that might result from extraction? Who should decide this, for example through processes of FPIC now required for projects involving indigenous communities? As pointed out by Torres in this collection, the geography of extraction often involves negative impacts at the local level, and capital accumulation in distant metropoles.

Assuming that resources are to be mined, the second question concerns which stakeholders should extract and/or process them. This is often posed as a legal issue (who has the 'right' under the existing legal framework) but is also an issue of values and ethics, informed by the often troubling history of mineral exploitation; as well as an issue of technical and institutional capacity: do local communities, cooperatives and small companies have the ability to extract strategic resources safely, without damaging the environment? Or can large-scale companies (most of them multinationals registered in the Global North) more efficiently make use of minerals, providing greater amounts of tax revenue to the state, which can be reinvested in public services? Can state agencies or NGOs enable communities to build their own capacities to manage and improve extractive processes, or are such initiatives merely tokens, to distract from larger patterns of dispossession of communities in favour of industrial mining companies?

These are inherently political questions, of course (though with 'technical' aspects) and the ways in which the questions are posed, and answered, depends on the prevailing political climate, whether at the local, national, or global level. It becomes important to examine the political economy of decision-making – that is, the overt and covert relationships between politicians, policy-makers, lobbyists, donors, business people, leaders of cooperatives or associations, and members of regulatory institutions charged with monitoring, evaluating or certifying resource use. Where politicians and policy-makers have vested personal interests, for example, due to a weak regulatory environment, or failure to systematically enforce existing laws, these questions are likely to be answered in ways that benefit them or their political allies and business partners. While in most cases the policy framework has been developed over long periods (as described in Campbell's paper in this collection), in some cases – following landslide election victories, for example, or in the aftermath of political or economic crisis – it may be possible for governments to make policy changes very rapidly, without popular consultation.[6]

These questions resonate with ongoing debates and policy processes in the forestry sector and wildlife conservation sphere, to give two examples, where models for co-management by communities and state agencies are common, but controversial. While they still have considerable support on the basis that they provide local people with some decision-making power and material benefits, they are often criticised as a fig leaf for state control; or because they place extra burdens on communities while providing few local benefits; or on the basis that they sometimes become corrupt; or that they are ineffective in controlling environmental degradation.[7] Controversies over community roles in natural resource management have become fiercer since the start of the reducing emissions from deforestation and forest degradation (REDD+) process and the establishment of a global market in carbon credits; which is the most visible of many schemes involving the commodification of 'environmental services' and resources.[8]

There are also strong similarities with the ways in which multinational companies have acquired large areas of farmland across the globe, especially since the food price crisis of 2007/2008; whereas in past decades, agrarian reform has often been characterized by a 'land to the tiller' approach (ie local people should be granted rights to own resources based on their historical use of them), the so-called global land grab was sometimes justified by companies and governments on the basis that communities did not have the capacity to produce high yields from farmland (with implications for food security and national economic growth). Alternatively, like 'green grabbing' in the name of conservation, acquisition of fertile land was based on the assumption that large areas of land – particularly in Africa – are 'vacant' and uninhabited.[9] When policy-makers try to determine whether there are 'local' communities with claims to resources, these discussions may aggravate existing tensions over territorial claims of self-defined 'autochtone' and 'migrant' communities.[10] The shift towards multinational acquisition of land was also possible due to decades of legal and policy reforms aimed at creating liberal land markets, similar to those described by Campbell in this collection.[11] These similarities, then, remind us that any discussion of property rights issues in the ASM sector should be informed by broader patterns in contemporary 'international development', such as the prevalence of market-based approaches, public–private partnerships, and the dominance of very large organisations (whether for-profit or non-profit) or coalitions of actors which are able to achieve 'impact at scale'.

The theme of property rights and ASM is not a new one. Strengthening the rights to land and resources of communities and indigenous populations was recommended more than a decade ago by *Striking a better balance: the World Bank Group and extractive industries*, the final report of the extractive industries review.[12] A more recent UN and African Union (AU) report notes that even when rights to access land are granted to artisanal and small-scale miners, these rights are rarely of sufficient duration to be deemed 'secure' rights for mining.[13] Despite these reports, 'the relationship between ASM and surface land tenure arrangements remains under-researched',[14] and this theme is particularly important today. It is a period of rapid change in the mining industry. For example, many mining companies have, for the first time, committed themselves to FPIC approaches[15] (though FPIC is still highly problematic and contested) and there has been a noticeable increase in multi-actor efforts to implement and enforce mineral certification schemes in areas where mining is particularly associated with human rights abuses and/or conflict. For example, following mainstream media coverage of the 'conflict diamonds' phenomenon – whereby diamond production directly or indirectly finances armed violence – the Kimberley Process (KP) was established in 2003. The KP is a voluntary international certification scheme for rough diamonds, incorporating diamond-producing states, diamond trading companies and civil society organisations. It enjoys support from the UN Security and includes 54 countries around the world. However, the scheme has been criticised by non-governmental organisations such as Global Witness and Partnership Africa Canada – who have ceased their involvement in the scheme – for weak monitoring and enforcement and for having an overly restrictive definition of conflict diamonds.[16]

In addition, the International Conference of the Great Lakes Region (ICGLR), in collaboration with other actors, has put in place a Regional Certification Mechanism to certify that production and trade in tin, tantalum, tungsten and gold in the Great Lakes Region do not contribute to conflict or serious human rights abuses.[17] In order to be effective, this certification scheme must be supported by strong mineral tracking mechanisms, regular third-party audits, and other ongoing verification processes, and it has aligned itself with private sector efforts, and is compliant with the requirements of the US Dodds–Frank Act. The first African Regional Forum on Business and Human Rights which was held in September 2014 also involved discussions of due diligence in the extractive industries.[18] While these initiatives provide opportunities for reform, they are to some extent embedded in the contentious 'conflict minerals' conceptual framework, which has been vigorously attacked by critical scholars, including contributors to the current collection.[19] For example, as pointed out by Global Witness and others, the 'conflict minerals' definition used by the KP only concerns minerals that finance rebel groups; not broader links to state-sponsored violence, or human rights abuses related to population displacement and other extreme socio-economic injustices.[20] The intense publicity around conflict minerals in the DRC, and the US legislation (as well as the related temporary DRC government ban on mineral production in the Kivu and Maniema Provinces in 2010), have been criticised as distracting attention from broader governance issues, and for having a negative impacts on the livelihoods of millions of artisanal miners and their families.[21]

Property rights and mining

From a broader perspective of property rights and globalisation, rapid changes to property rights regimes are now occurring around the world, due to internal and external pressures on Southern governments to facilitate foreign direct investment (FDI); as well as urbanisation, environmental change, population growth, land and agricultural reforms, and economic expansion in the Global South. This is part of the broader neoliberal policy agenda of many of the most economically and politically powerful countries. The individualisation and commodification of land, minerals and other resources can be seen as a form of accumulation by dispossession, a version of what Marx called 'Primitive accumulation'.[22] From this perspective, 'the advanced capitalist states have successfully sought out new spaces of accumulation, or dramatically reorganised older colonial ones in their own interests, to absorb the North's surpluses and boost flagging profitability'.[23] Finance for mining operations in the Global South has increased when financial crises in the Global North, elements of a cyclical crisis of accumulation, has made the South a more secure and profitable investment. However, while recognising the very uneven power dynamics in international relations, we should not see Southern governments as powerless. Most governments, over the past 25 years, have 'actively constructed a set of legal, fiscal and political incentives to attract FDI towards the development of internal resources.'[24] Much of what has been public – common property resources like forests and pastures – is becoming enclosed, privatised, and commodified.[25] Nevertheless, there are counter-tendencies. As noted above, concerns around the so-called global land grab resulted in new normative frameworks that seek to guide 'investment'.[26] Protection of 'community' or customarily owned land and resources is increasingly promoted through regional efforts and United Nations programmes.[27] Some governments are trying to put in place land and property rights systems that provide customary rights holders with improved land tenure security, without simply converting customary rights to formal title or lease.

From a legal perspective, in most countries around the world except the United States, minerals are owned and licensed by the state, regardless of who claims ownership of the surface resources.[28] In other words, there is a separation between surface rights and sub-surface rights. In many countries, mining concessions can only be issued to large-scale, industrial actors. In other cases, specific areas are set aside for ASM, but these are rarely very valuable deposits, and the regulatory mechanisms involved can be difficult to navigate.

People claiming land rights in areas under a mining concession may be prevented from carrying out certain activities, depending on the location and scale of exploration or exploitation of minerals. Parts of the concession are fenced off to prevent 'trespassing'. In the words of one scholar, 'mining investment in most instances simply cannot proceed without a community – often indigenous – being dispossessed of their land, natural resources and livelihoods'.[29] 'Dispossession', argues one critic, is 'at the centre of the activities of the mining industry'.[30] This dispossession, inequities in distribution of profits from the mining sector, and marginalisation of artisanal miners, has led directly to violence in many places, including Colombia,[31] Tanzania,[32] and the Democratic Republic of Congo (DRC)[33] to name just a few. In cases where people are displaced from their homes, there are international global norms and legal standards that should guarantee them fair, timely compensation and due process[34]; though these are not always followed in practice.

Typically, the concession is seen as the possession of the mining company, for the duration of the permit. This is a simplistic framework, however. Just as rights to land are often understood as a 'bundle of rights' of different kinds,[35] in the mining sector, various kinds of rights can also be identified, in theory and in practice. These include, for example, rights to conduct geological exploration activities in a particular area; rights to remove minerals from the soil; rights to possess mineralised materials on mining sites; rights to transport minerals; rights to sell minerals; rights to use a form of title to a mineral-rich area as collateral for credit, and the rights to control the mineral production from a particular pit or mining site. Rights to minerals should therefore be seen as a 'bundle of rights'. Though more powerful companies typically attempt to control the entire bundle f rights, they may be divided up between different actors, often through informal arrangements, and there may be ways in which holders of different surface and sub-surface rights to the same area can coexist. In South

Africa, for example, gold mining generally takes place underground and can thus be conducted underneath large-scale maize farms.[36]

The formal legal distinction between surface and sub-surface rights is rarely reflected in customary practices. Local people may demand payments from mining crews or companies on the basis that they are the traditional landowners. In some countries, such as Ghana, customary land ownership is legally recognised, so that customary land rights holders can legally claim land rents and a portion of mining royalties. The importance of custom cannot be underestimated, particularly because much artisanal mining is found in relatively remote areas, where land is not registered under state-managed systems. For some analysts, the policy challenge is to marry statutory and customary tenure systems in ways suited to the cultural and economic conditions of particular countries.[37]

Scholars should therefore beware of drawing a clear line between 'custom' and 'state' systems; due to local hybridity, a typical mining area may be best characterised as a 'multi-layered "state arena" that has become increasingly fragmented as a result of processes of decentralization and privatization'.[38] For many analysts, one of the main challenges is the legitimacy and capacity of the state to act as an arbiter in the public interest. However, some scholars refuse to subscribe to a simple definition of state sovereignty over mineral resources, focusing instead on 'the multiple social relations and political forces struggling over resources and territory',[39] and the tensions that are introduced when the state – which claims exclusive rights over sub-surface resources – lacks the technical, financial or organisational capacity to mine, process and trade them on the global market.

As is evident in this collection, institutional, economic and legal analyses only capture some aspects of land and mineral rights; they encompass diverse elements and are viewed very differently in different disciplines. Land and property is also understood according to value systems particular to communities, identity groups and individuals. For example, land is significant as an area where political authority is expressed and taxes may be raised (the concept of 'territory'), as a means through which families and individuals maintain socio-cultural status, and also as a source of feelings of historic individual or communal 'belonging'; dependence upon land and natural resources is key to the notion of indigeneity.[40] Inhabitants of settlements in and around mineral-rich areas may have diverse sets of opinions about the 'value' of land and minerals, and these values are not mutually exclusive. While one author correctly argues that most land-related conflicts around mining are caused by, 'mine management neglecting the needs of indigenous people who have cultural attachments to their environment',[41] an individual may have strong feelings both about the need to preserve the socio-ecological integrity of a particular area; but also desire economic development. Perceptions of 'development' are dynamic and contingent on complex socio-political factors at different levels. By considering power relations and discursive debates over resources, scholars may reconceptualise a 'commodity' as a dynamic field of struggle, or a mining concession, for example, as, 'contested space characterized by accelerated change and an increased density of activity and claims, where power is deployed in various forms'.[42]

The artisanal miner as subject

This idea of mining site as field of power reminds us of the importance of structural elements to our analysis. There is an influential trope in much of the writing, particularly the policy literature but also some academic work, which characterises ASM as 'anarchic' or 'chaotic'. In fact, such mining is characterised by some level of organisation, negotiation and planning.[43] The level and nature of organisation will depend on the geological, institutional, legal and other characteristics of the environment. For example, in Brazil and Venezuela, miners organise themselves into artisanal diamond mining cooperatives primarily to secure access to mineralised land. Once access is secured, cooperatives cease functioning and mining crews work separately.[44] In Mozambique, the level or organisation depends partly on where the gold is located, with different levels of organisation necessary to mine 'placer' gold deposits (often found near the surface) and 'reef' deposits (found deeper underground).[45]

The term ASM has also over time come to be associated, especially in the policy-oriented literature, with a host of social and environmental ills including pollution (especially the use of mercury and other chemicals), failure to rehabilitate mine sites, dangerous labour conditions, child labour, various forms of sexual and economic exploitation of women and girls, drug use, and in some countries, smuggling of minerals to avoid government taxes, and links to organised crime and armed groups.[46] Disputes and struggles over rights to minerals and related resources (land, water, etc.) are not unique to low-income countries, but have been common throughout the history of mining in most parts of the world.[47] However, it is notable that ASM is not framed in the same way as some of the historical forms of informal mining in the west; independent miners in low-income countries are rarely seen as 'pioneers' for example.[48] It has instead been described, in some work, as a 'primitive' form of resource extraction.[49] Some authors identify the term ASM, and its various associations, with a thinly veiled discourse of racism and primitivism, according to which such miners are not only less skilled, but also less rational – in other words, less human – than those involved in industrial mining.[50] For this reason, some authors refuse to use the term 'ASM', referring instead to 'citizen miners' or 'independent miners'.[51]

Miners are typically criminalised for working outside of state-managed systems; but are rarely provided with incentives to gain formal legal recognition, due to a number of factors including a shortage of mineralised lands which are legally available to ASM actors, expensive licensing fees, corruption and bureaucratic permit systems causing long delays.[52] In Ghana, for example, where an estimated 12% of the total land area is currently under mining concessions,[53] much mineralised land has been granted to industrial operators, or is categorised as forest reserve.[54] Experience shows that approaches which rely primarily on 'legislating' or 'policing' to ensure that miners follow the regulations, rather than using incentives, are unlikely to succeed. Sometimes property rights are put forward as a 'carrot' to encourage artisanal miners to formalise and 'legalize' their production[55]; however, one of the reasons for taking a critical approach to land and property rights is that it allows scholars to take an less instrumental approach.

Contributions to this collection

The articles in this collection use insights from different theoretical schools to understand this most interdisciplinary of topics. Campbell,[56] in her article on regulatory frameworks for ASM, distils findings from fifteen years of research, and urges scholars to re-consider the conceptual lenses through which we view the phenomenon of informal mining. Basing her analysis on insights from legal pluralism and heterodox international political economy, she critiques the neoliberal model of 'liberalization' on the basis that it has resulted in governments handing over much responsibility for minerals (and related livelihoods) to private companies, privileged foreign 'investment' to the detriment of building local and national capacities, treated negative social and environmental impacts as peripheral to economic priorities, and treated mining as a 'separate' activity from other aspects of rural life. In contrast to these dominant approaches, the article discusses the current Senegalese regulatory framework as a promising alternative. Campbell's paper ensures that this collection considers the issue of property rights from within a comprehensive and critical analysis of the broader governance context.

In the next article, Hatcher[57] discusses the case of Mongolia, which has undergone a rapid and dramatic process of mining sector expansion. Hatcher uses a political economy analysis to discuss the dichotomy between the industrial mining sector – which includes some of the largest coal, copper and gold mines in the world – and the ASM sector, which employs some 20% of the rural workforce, but has been, until recently, forced into a situation of illegality. She notes that the large profits from the industrial mining sector have had little impact on poverty rates in Mongolia, and shows how artisanal miners, known locally as 'ninjas', have been marginalised not just through regulation of the mining sector but also because of deteriorating relationships with pastoralists and other segments of the rural population. Hatcher's contribution is useful not just as a comprehensive assessment of a hitherto poorly documented country case study, but also as evidence of the potentials of combining fine-grained analysis of institutional dynamics with a broader macro-economic perspective (which resonates with Campbell's). She points to the benefits of legalisation of the sector, particularly in terms of improved access to social services for miners and their families.

The third article, by Katz-Lavigne,[58] provides an overview of classical theories on property rights, especially the evolutionary theories of property rights that assume particular rationalities of privatisation. Critiquing these theories, and focusing particularly on issues of economic efficiency and distributional concerns, she examines the role of the state in property rights enforcement in areas of 'limited statehood'. Using the examples of industrial mining in the DRC and Zambia (and associated disputes with ASM actors), Katz-Lavigne uses the concept of 'enforcement coalitions' to examine the ways in which states, companies, local communities, and other actors interact, often via (fragile) 'corporate social responsibility'-type commitments. She also looks notes the differential gender impacts of dispossession of local communities. This contribution is particularly interesting for its constructively critical engagement with classical theory.

In his article, Mnwana[59] critically examines the decentralised governance regime in South Africa, paying particular attention to the role of chiefs as power-brokers. In an analysis that will resonate with scholars of African land rights and governance regimes,[60] Mnwana identifies particular interpretations of custom – based on 'crystallizations' of what were, in practice, somewhat flexible regulations, accountable to local people – that have centred control over community mineral resources (platinum in this case) in the institution of the chief. By documenting community efforts to challenge the chief's powers – which often seem to be abused for private profit – Mnwana affirms the agency of community members. However, the article also argues that the formal legal system has become instrumentalised as a mechanism 'for chiefs to suppress opposition'. Mnwana makes an important contribution to our understanding of the socio-political impacts of industrial mining when he concludes that struggles over 'customary' powers may threaten the coherence of the higher level legal notion of 'tribal community'.

Vélez-Torres[61] examines the ways in which particular forms of ASM associated with rural Afro-Colombian communities have become an intrinsic element of these communities' positions and struggles against what she terms a 'double-faced' phenomenon of extractive mining, conducted by multinational companies on the one hand, and smaller scale illegal miners, using mechanised backhoes, on the other. Vélez-Torres highlights the aspect of 'land' as territory, and also shows that particular forms of small-scale mining are intrinsic to local identities. Without essentialising 'communities', she argues that 'the cosmo-visions and cosmo-actions of communities' are not just part of the context of conflict, but are also 'central to resistance'. While acknowledging the progressive legal frameworks that the Colombian government has established, she argues that they have not yet been fully implemented and highlights the efforts of Afro-descendants to struggle for their rights through multiple strategies, including 'radical mobilization' as well as legal contests in the courts.

Geenen and Claessens[62] speak to the literature on 'conflict minerals', which they characterise as 'ahistorical and reductionist'. Their article uses the situation in Mukungwe, Eastern DRC, to elaborate the various different ways that actors attempt to control surface and sub-surface resources. Using concepts derived from Marxist political ecology, Geenen and Claessens highlight 'relational mechanisms' through which elites and members of state agencies, for example, enter into (often temporary) alliances and networks, as well as a set of strategies including legitimisation and legalisation of informal activities, using legal action against opponents, creating new institutional forms, and mobilising violence. Their analysis manages to show how actors seek control over mining areas through territorialisation ('controlling space as well as people in that space'), as well as through means more associated with land and minerals as commodities. As such, it is an interesting case to compare with that of Colombia, discussed above.

Finally, van de Camp[63] uses ethnographic methods, and a concept of mines as complex socio-ecological systems, to develop a detailed description and analysis of the ways in which miners negotiate access to minerals across 'three dimensions': the socio-political dimensions at the surface (two dimensions), and the geological aspects underground (a third dimension). By describing the geological details of gold mining in Northern Ghana, van de Camp provides a very rich image of everyday mining practices. Negotiations over access to subsurface resources involve not only political and financial manoeuvrings, but also engagement with the cosmological aspects of mining; the customary beliefs and values that often frame the exploitation of minerals. For this reason, van de Camp advises, we should consider gold to be a 'substance' (to which people apply social, cultural, cosmological, ecological and geological meanings) and not just a 'resource'.

These articles are notable for their clear-eyed assessment of the various benefits and risks associated with the ASM sector, their recognition of the agency of artisanal miners and 'local communities' within the uneven hierarchies in which they are embedded, and their acknowledgement of the difficulties of state regulation of such a complex set of issues. Several contributors refer to similar frameworks (broadly, critical understandings of property access which are attuned to multi-scale power dynamics) but use them to develop somewhat different analytical tools. These attributes, as well as a common characteristic of identifying general findings from particular case studies, make this collection a valuable addition to the literature on property rights and mining.

Disclosure statement

No potential conflict of interest was reported by the author.

Funding

This research was supported in part by a post-doctoral fellowship from the Netherlands Academy on Land Governance and Food Security (LANDac), with funding from CIFOR [project funding WS.320013.1].

Acknowledgements

The author would like to thank LANDac personnel who helped to organise the LANDac conference, Utrecht University, 2015, from which this collection originates.

Notes

1. Buxton, *Responding to the Challenge*; and Ruzvidzo, *African Women*.
2. Nyame and Blocker, "Influence of Land Tenure"; and Lahiri-Dutt, "The Feminisation."
3. Butler, *Colonial Extractions*, 217.
4. Zoomers, "Globalization."
5. UN Food and Agriculture Organization (FAO), *Voluntary Guidelines*.
6. Klein, *The Shock Doctrine*.
7. Cronkleton et al., "Co-management in Community Forestry"; Horwich and Lyon, "Community Conservation"; and Goldman, "Partitioned Nature."
8. Leach and Scoones, *Carbon Conflicts*.
9. Zoomers, "Globalization."
10. Geschiere, *Perils of Belonging*.
11. Zoomers, "Globalization."
12. Cited in Campbell, *Mining in Africa*, 15.
13. Cited in Campbell, this issue.
14. Verbrugge et al., "Min(d)ing the Land," 50.
15. Oxfam International, *Community Consent Index 2015*.
16. Global Witness, *Global Witness Leaves Kimberely Process*.
17. ICGLR, *Protocol against the Illegal Exploitation*.
18. Taka, *Critical Analysis of Human Rights*.
19. See eg Johnson, *No Kivu, No Conflict?* and Seay, *What's Wrong with the Dodd-Frank*, cited in Geenen and Claessens, this issue.
20. Sharife and Grobler, "Kimberley's Illicit Process."
21. See eg Seay, *The Dodd-Franks Act*.

22. Marx, *Capital*, 714; and Harvey, *A Brief History of Neoliberalism*, 159.
23. Gordon and Webber, "Imperialism and Resistance," 65–6.
24. Emel et al., "Extracting Sovereignty," 72.
25. See eg Mansfield, "Privatization"; Maurer and Schwab, *Accelerating Possession*; Anseeuw et al., *Land Rights*; and Cotula, *The International Political Economy.*
26. See eg UN Food and Agriculture Organization (FAO), *Voluntary Guidelines.*
27. See eg AUC-ECA-AfDB Consortium, *Land Policy in Africa.*
28. Woertz, "Mining Strategies in the Middle East"; and Emel et al., "Extracting Sovereignty."
29. Gordon and Webber, "Imperialism and Resistance," 67.
30. Ibid.
31. Ibid., 78.
32. Emel et al., "Extracting Sovereignty," 77.
33. Geeneen and Claessens, this issue; and Claessens et al., *Competition over Soil and Subsoil.*
34. For an overview of international law on property rights, especially in the context of forced or market-based eviction, see Leckie and Huggins, *Conflict and Housing, Land*, 53–91.
35. Di Robilant, "Property: A Bundle of Sticks"; and Freudenberger et al., *Property Rights and Artisanal Mining*.
36. Verbrugge et al., "Min(d)ing the Land"; and Pijpers, "Crops and Carats," 35.
37. Freudenberger et al., *Property Rights and Artisanal Mining*.
38. Verbrugge et al., "Min(d)ing the Land," 59.
39. Emel et al., "Extracting Sovereignty," 72.
40. Marschke et al., "Engaging Indigeneity in Development."
41. Hilson, "An Overview of Land use Conflicts," 72.
42. Pijpers, "Crops and Carats," 35.
43. Dondeyne and Ndunguru, "Artisanal Gold Mining and Rural Development," 122; and Butler, *Colonial Extractions*, 218.
44. Blore, *Trading Tenure for Formalization*, 49.
45. Dondeyne and Ndunguru, "Artisanal Gold Mining and Rural Development," 123–4.
46. See eg World Bank, *Strategy for African Mining*, cited in Butler, *Colonial Extractions*, 218; Freudenberger et al., *Property Rights and Artisanal Mining*, 2; Hilson and Gatsinzi, "A Rocky Road Ahead," 3; and Verbrugge et al., "Min(d)ing the Land," 51.
47. Freudenberger et al., *Property Rights and Artisanal Mining*. One memorable cinematic depiction of social tensions around the extractive industries in Western history is the 2007 film 'There will be Blood' (Dir. Paul Thomas Anderson), which shows the controversies over land and oil rights in Southern California, USA, in the late nineteenth century.
48. Butler, *Colonial Extractions.*
49. Barry, *Regularizing Informal Mining. Cited in* Butler, *Colonial Extractions.*
50. Butler, *Colonial Extractions*, 217–8.
51. Lissu, Tindu (no specific publication cited) cited in *Cited in* Butler, *Colonial Extractions, 213.*
52. Hilson and Gatsinzi, "A Rocky Road Ahead," 3.
53. Banchirigah and Hilson, *De-agrarianization, Re-agrarianization*.
54. Akabzaa, *Mining in Ghana.*
55. Blore, *Trading Tenure for Formalisation.*
56. Campbell, "Revisiting Artisanal and Small."
57. Hatcher, "The Politics of Artisanal."
58. Katz-Lavigne, "Property Rights and Large-scale."
59. Mnwana, "Fragmented 'Custom' and Fractured 'Community'."
60. Pottier, "'Customary land tenure'."
61. Vélez-Torres, "Disputes Over Gold Mining."
62. Geenen and Claessens, "Different Faces of Access."
63. van de Camp, "Artisanal Gold Mining in Kejetia."

ORCiD

Chris Huggins http://orcid.org/0000-0002-1318-9194

Bibliography

Akabzaa, T. "Mining in Ghana: Implications for National Economic Development and Poverty Reduction." In *Mining in Africa: Regulation and Development*, edited by Bonnie Campbell, 25–65. Uppsala: The Nordic Africa Institute, 2009.

Akabzaa, Thomas. "Mining in Ghana: Implications for National Economic Development and Poverty Reduction." In *Mining and Development in Africa*, edited by Bonnie Campbell, 25–65. Montreal: Pluto Publishers, 2009.

Anseeuw, Ward, Liz Alden Wily, Lorenzo Cotula, and Michael Taylor. *Land Rights and the Rush for Land: Findings of the Global Commercial Pressures on Land Research Project*. Rome: International Land Coalition, 2011.

AUC-ECA-AfDB Consortium. *Land Policy in Africa: A Framework to Strengthen Land Rights, Enhance Productivity and Secure Livelihoods*. Addis Ababa: AUC-ECA-AfDB Consortium, 2010.

Banchirigah, Sadia Mohammed, and Gavin Hilson. "De-agrarianization, Re-agrarianization and Local Economic Development: Re-orientating Livelihoods in African Artisanal Mining Communities." *Policy Sciences* 43, (2010): 157–180.

Barry, Mamadou, ed. *Regularizing Informal Mining: A Summary of the Proceedings of the International Roundtable on Artisanal Mining*. Washington, DC: World Bank, 1996.

Blore, Shawn. "Trading Tenure for Formalisation: A New Approach to the Licensing of Artisanal Diamond Mining in the DRC." In *A Farmer's Best Friend? Artisanal Diamond Mining and Rural Change in West and Central Africa*, edited by S. Van Bockstael and K. Vlassenroot, 33–68. Gent: Egmont Institute, Academia Press, 2012.

Buxton, Abbi. *Responding to the Challenge of Artisanal and Small-scale mining. How Can Knowledge Networks Help?* London: IIED, 2013.

van de Camp, Esther. "Artisanal Gold Mining in Kejetia (Tongo, Northern Ghana): A Three-dimensional Perspective." *Third World Thematics* 1, (2016): 267–283. doi:10.1080/23802014.2016.1229132.

Campbell, Bonnie. "Introduction." In *Mining in Africa: Regulation and Development*, edited by Bonnie Campbell, 1–24. Uppsala: The Nordic Africa Institute, 2009.

Campbell, Bonnie. "Revisiting the Interconnections Between Research Strategies and Policy Proposals: Reflections from the Artisanal and Small-Scale Mining Sector in Africa." *Third World Thematics*, (2016): 165–183. doi:10.1080/23802014.2016.1226145.

Claessens, Klara, Emery Muginga, and An Ansoms. "Competition over Soil and Subsoil: Land Grabbing by Local Elites in South Kivu (DRC)." In *Losing Your Land: Dispossession in the Great Lakes*, edited by A. Ansoms and T. Hilhorst, 82–102. Martlesham: James Currey, 2014

Cotula, Lorenzo. "The International Political Economy of the Global Land Rush: A Critical Appraisal of Trends, Scale, Geography and Drivers." *The Journal of Peasant Studies* 39, no. 3–4 (2012): 649–680.

Cronkleton, P., J. M. Pulhin, and S. Saigal. "Co-management in Community Forestry: How the Partial Devolution of Management Rights Creates Challenges for Forest Communities." *Conservation and Society* 10, (2012): 91–102.

Di Robilant, Anna. "Property: A Bundle of Sticks or a Tree?" *Vanderbilt Law Review* 66, no. 3 (2013): 870–932.

Dondeyne, Stefaan, and Eduardo Ndunguru. "Artisanal Gold Mining and Rural Development Policies in Mozambique: Perspectives for the Future." *Futures* 62, (2014): 120–127.

Emel, Jody, Matthew T. Huber, and Madoshi H. Makene. "Extracting Sovereignty: Capital, Territory, and Gold Mining in Tanzania." *Political Geography* 30, (2011): 70–79.

Extractive Industries Review. *Striking a Better Balance: The World Bank Group and Extractive Industries*. Jakarta: World Bank Group and Extractive Industries, 2003.

Freudenberger, Mark, Saleem Ali, and Tim Fella. *Property Rights and Artisanal Mining: Clarifying and Strengthening Rights: Options for Policy Makers*. Property Rights and Resource Governance Briefing Paper # 14. Washington, DC: USAID, 2011.

Freudenberger, Mark with Bocar Thiam, Catherine Picard, Sabine Jiekak, and Terah DeJong. "Property Rights Conundrums in Artisanal and Small-scale Mining: Experiences from USAID Projects in Côte D'ivoire, Guinea, and the Democratic Republic Of Congo." Paper prepared for presentation at the 2015 World Bank Conference on Land and Poverty, The World Bank, Washington, DC, March 23–27, 2015.

Geenen, Sara, and Klara Claessens. "Different Faces of Access Control in a Congolese Gold Mine." *Third World Thematics*, (2016): 249–266. doi:10.1080/23802014.2016.1180544.

Geschiere, Peter. *The Perils of Belonging: Autochthony, Citizenship and Exclusion in Africa and Europe*. Chicago: University of Chicago Press, 2009.

Global Witness. *Global Witness Leaves Kimberley Process, Calls for Diamond Trade to be Held Accountable*. Press Release, December 2, 2011.

Gneiting, U., T. Bruno-Van Vijfeijken, H. P. Schmitz, and R. Gomez. "Setting Higher Goals: Rights and Development Trade-offs and Challenges in Implementing a Rights-based Approach to Development." *Monday Development*, December 2009.

Goldman, Mara. "Partitioned Nature, Privileged Knowledge: Community Based Conservation in Tanzania." *Development and Change*. 34, no. 5 (2003): 833–862.

Gordon, Todd, and Fery R. Webber. "Imperialism and Resistance: Canadian Mining Companies in Latin America." *Third World Quarterly* 29, no. 1 (2008): 63–87.

Harvey, David. *A Brief History of Neoliberalism*. New York: Oxford University Press, 2005.

Hatcher, Pascale. "The Politics of Artisanal and Small-scale Mining in Mongolia." *Third World Thematics*, (2016): 184–201. doi:10.1080/23802014.2016.1231592.

Hilson, Gavin. "An Overview of Land Use Conflicts in Mining Communities." *Land Use Policy* 19, (2002): 65–73.

Hilson, Gavin, and Angelique Gatsinzi. "A Rocky Road Ahead? Critical Reflections on the Futures of Small-scale Mining in Sub-Saharan Africa." *Futures* 62, (2014): 1–9.

Horwich, Robert, and Jonathan Lyon. "Community Conservation: Practitioners' Answer to Critics." *Oryx* 41, no. 03 (2007): 376–385.

International Conference on the Great Lakes Region (ICGLR). *ICGLR Regional Certification Mechanism (RCM) – Certification Manual*. Kampala: ICGLR, n.d.

International Conference on the Great Lakes Region (ICGLR). *Protocol against the Illegal Exploitation of Natural Resources*. Kampala: ICGLR, November 30, 2006.

Johnson, Dominic. *No Kivu, No Conflict? The Misguided Struggle Against Conflict Minerals in the DRC*. Goma: Pole Institute, 2013.

Katz-Lavigne, Sarah. "Property Rights and Large-scale Mining: Overlapping Claims at and Around Mining Sites at the Democratic Republic of Congo and Zambia." *Third World Thematics*, (2016): 202–217. doi:10.1080/23802014.2016.1196604.

Klein, Naomi. *The Shock Doctrine: The Rise of Disaster Capitalism*. Toronto: Alfred A. Knopf Canada, 2007.

Lahiri-Dutt, Kuntala. "The Feminisation of Mining." *Geography Compass* 9, no. 9 (2015): 523–541.

Leach, Melissa, and Ian Scoones, eds. *Carbon Conflicts and Forest Landscapes in Africa*. London: Routledge.

Leckie, Scott, and Chris Huggins. *Housing, Land, and Property Rights and Conflict: A Handbook on Issues, Frameworks and Solutions*. Cambridge: Cambridge University Press, 2011.

Li, Tania M. "Practices of Assemblage and Community Forest Management." *Economy and Society* 36, (2007): 263–293.

Mansfield, Becky. "Privatization: Property and the Remaking of Nature-Society Relations. Introduction to the Special Issue." *Antipode* 39, no. 3 (2007): 393–405.

Marschke, M., D. Szablowski, and P. Vandergeest. "Engaging Indigeneity in Development Policy." *Development Policy Review* 26, no. 4 (2008): 483–500.

Marx, Karl. *Capital. Volume I*. New York: International Publishers, 1967.

Matthysen, K. *Review of the Burundian Artisanal Gold Mining Sector*. Brussels: IPIS, 2015.

Maurer, Bill, and Gabriele Schwab. *Accelerating Possession: Global Futures of Property and Personhood*. New York: Columbia University Press, 2006.

Mnwana, Sonwabile. "Fragmented 'Custom' and Fractured 'Community': Mining and Property Struggles on the Platinum Belt, South Africa." *Third World Thematics*, (2016): 218–234. doi:10.1080/23802014.2016.1202776.

Mommer, B. *Global Oil and the Nation State*. Oxford: Oxford University Press, 2002.

Nyame, Frank K., and Joseph Blocher. "Influence of Land Tenure Practices on Artisanal Mining Activity in Ghana." *Resources Policy* 35, (2010): 47–53.

Oxfam International. *Community Consent Index 2015: Oil, Gas, and Mining Company Public Positions on Free, Prior, and Informed Consent*. Briefing Paper. Oxford: Oxfam International, July 3, 2015.

Park, Clara Mi Young, and Ben Julia White. "We are Not all the Same: Taking Gender Seriously in Food Sovereignty Discourse." *Third World Quarterly* 36, no. 3 (2015): 584–599.

Patel, Reena. "Hindu Women's Property Rights in India: A Critical Appraisal." *Third World Quarterly* 27, no. 7 (2006): 1255–1268.

Pijpers, Robert. "Crops and Carats: Exploring the Interconnectedness of Mining and Agriculture in Sub-Saharan Africa." *Futures* 62, (2014): 32–39.

Pottier, Johan. "'Customary land tenure' in Sub-Saharan Africa Today: Meanings and contexts." In *From the Ground Up: Land Rights, Conflict and Peace in Sub-Saharan Africa*, edited by Chris Huggins and Jenny Clover, 55–75. Pretoria: Institute for Security Studies, 2005.

Ruzvidzo, T. Mining Must be Gender-inclusive, in African Minerals Development Centre. *African Women in Artisanal and Small-scale Mining*. Addis Ababa: ADMC/African Union, 2015.

Seay, Laura. *What's Wrong with the Dodd-Frank 1502? Conflict Minerals, Civilian Livelihoods, and the Unintended Consequences of Western Advocacy*. Working Paper 284. Washington, DC: Center for Global Development, 2012.

Sharife, Khadija, and John Grobler. "Kimberley's Illicit Process." *World Policy Journal* Winter 2013/2014, (2013) : 65–77.

Taka, Miho. *A Critical Analysis of Human Rights Due Diligence Processes in Mineral Supply Chains: Conflict Minerals in the DRC*. Occasional Paper 208. South African Institute of International Affairs, 2014.

UNECA & AU. *Minerals and Africa's Development. A Report of the International Study Group on Africa's Mineral Regimes*. Addis Ababa: UNECA/African Union, 2011.

UN Food and Agriculture Organization (FAO). *Voluntary Guidelines on the Responsible Governance of Tenure of Land, Fisheries and Forests in the Context of National Food Security*, 2012. Accessed November 13, 2015. http://www.refworld.org/docid/5322b79e4.html

Vélez-Torres, Irene. "Disputes Over Gold Mining and Dispossession of Local Afrodescendant Communities from the Alto Cauca, Colombia." *Third World Thematics*, (2016): 235–248 doi:10.1080/23802014.2016.1229131.

Verbrugge, Boris, Jeroen Cuvelier, and Steven Van Bockstael. "Min(d)ing the Land: The Relationship between Artisanal and Small-scale Mining and Surface Land Arrangements in the Southern Philippines, Eastern DRC and Liberia." *Journal of Rural Studies* 37, (2015): 50–60.

Woertz, E. "Mining Strategies in the Middle East and North Africa." *Third World Quarterly* 35, no. 6 (2014): 939–957.

World Bank. *Strategy for African Mining*. World Bank Technical Paper Number 181. Washington, DC: World Bank, 1992.

Zoomers, Annelies. "Globalization and the foreignisation of space: Seven processes driving the current global land grab." *The Journal of Peasant Studies* 37, no. 2 (2010): 429–447.

Revisiting the interconnections between research strategies and policy proposals: reflections from the artisanal and small-scale mining sector in Africa

Bonnie Campbell

ABSTRACT

The objective of this article is to explore the interconnections between the choice of conceptual and methodological frameworks and resulting institutional reform and policy processes and to do so from a broad developmental perspective. It builds on the analytical framework developed in past research which examined the links between issues of regulation and legitimacy in the African mining sector. It draws attention to the structural relations of power institutionalised and reproduced by past regulatory frameworks, in order to analyse how this heritage has conditioned approaches to artisanal and small-scale mining (ASM) and, in turn, policy proposals. The article uses ASM as an example to illustrate the need to rethink appropriate categories to capture shifting boundaries and to re-assess approaches as preconditions to renew institutional reform and policy processes from a country-specific and participative perspective.

Introduction

This article has as its objective to illustrate certain conceptual and methodological obstacles to reformulating policy proposals from a broader development perspective. It uses the artisanal and small-scale mining (ASM) sector to highlight these interconnections. The point of entry of the article is to suggest that analysis in this area needs to be contextualised within the specific historical settings of which ASM and the processes in which it is embedded are the expression. As with the issue of tracing revenue flows or understanding disappointing development impacts, in approaching issues regarding the artisanal and small-scale mining sector in Africa, particular attention needs to be given to the heritage of reforms which have redefined regulatory frameworks over the last 30 years. For it is this process of reform which gave rise to the extensive liberalisation of the mining sector, the introduction of the 'investment-led model' and the reconfiguration of the relations among actors involved in mining activities.

By way of a specific illustration of the manner in which a particular orientation was introduced under the auspices of the World Bank through the reform process of regulatory frameworks in the mining sector in Africa, one could cite the recommendations set out in the 1992 document *Strategy for African Mining*. These were informed not by the perspectives of those responsible for African administrations, of mine ministries or of local populations affected by mining activities, but by the answers to questionnaires provided by foreign companies concerning their needs.[1]

It can be shown that the far-reaching reforms, which have stretched over several decades, gave rise to what have been identified as several generations of mining regimes and the introduction of a specific 'investment-led' development agenda, premised on a series of norms or principles which merit close attention.

Five features of past approaches will serve to illustrate this:

(a) Past reforms and resulting institutional arrangements were informed by an approach which was rigorously mono-sectorial – that is concerned only the mining sector. As the last three decades demonstrate, because foreign companies are above all integrated into supply and market chains internationally, they have tended not to integrate mining operations into local economies.

(b) Institutional reforms have been undertaken overwhelmingly from the perspective of mining companies rather than that of promoting development strategies. As noted, reforms which resulted from the World Bank's important 1992 study *Strategy for African Mining*, were informed by the replies of major and junior companies to questionnaires sent to 80 international mining companies[2] as to what types of regimes might best suit their needs, as were the recommendations of a subsequent study undertaken for the Bank almost 10 years later.[3]

(c) The central contribution of mining activities to the local economy was suggested to be mining revenues and foreign exchange receipts rather than developing linkages.[4] Moreover, because African countries were perceived to be a high- or medium-risk investment, they were encouraged to be competitive in their taxes and incentives.

(d) Concerning the environmental consequences of mining activities, and subsequently, the social impacts which were in fact both much less developed dimensions than other aspects of regulatory frameworks, these impacts were seen as side effects, considered as secondary and marginal to the positive consequences of investments. It was assumed that they could be regulated by the introduction of voluntary performance standards for which responsibility was seen to rest above all with companies rather than being considered as issues which were clearly interrelated and integral parts of development strategies entailing overriding government responsibility.

(e) Above all else and most important was the stringent retrenchment of the state which took place, and with it the new delineation between public and private spheres of authority and the transfer of what were previously considered public responsibilities to private actors. The private sector was to become 'owner and operator' – the role of government was relegated to that of facilitator and promoter ('regulator and promoter').[5]

Africa's experience over the last 30 years has been a cumulative process of reform leading to several generations of increasingly liberalised mining regimes.[6] As various studies bring out, the model which has been introduced to liberalise the sector was to have very profound implications for existing institutional arrangements and political processes with important consequences for the legitimacy of the operations of mining companies.[7]

Research conducted on this process and its consequences has led to identifying a series of *norms* born out of economic liberalism on which the liberalisation reform process has rested and continues to rest. Interestingly, analysis of the Canadian mining experience suggests a strong correspondence between the norms informing the liberalisation process of mining regimes in Africa and the principle of *free mining* in Canada.[8] Among the norms informing this reform process are the following:[9]

(1) Priority given to the private sector for mining;
(2) Priority given to mining over other types of land use;
(3) Priority given to an exportable resource (gold) over other mineral resources;
(4) Priority given to the industrial sector over artisanal and small mines;
(5) Guarantees protecting mining rights.

The above norms and the structural relations of influence and power of which they are the expression have obviously been central to the determination of the manner in which issues concerning land governance and property rights in the sector interact. There now exists a vast literature devoted to the implications of this model and its far-reaching economic and socio-environmental impacts and human right violations. Much less attention appears to have been directed to the analysis of the manner in which these norms have informed the development agenda and resulting strategies and have impacted on our way of conceptualising specific areas of inquiry, as for example in the artisanal and small-scale mining sector. In order to contribute to reflection in this area, our analysis which remains very much at the exploratory stage will address the interconnections between three areas where further work would seem necessary:

(1) Rethinking appropriate categories to capture shifting boundaries;
(2) Revisiting approaches;
(3) Renewing institutional reform and policy processes.

It should be stressed that there exists a very rich body of literature that deals with the ASM-phenomenon and this article does not have the pretence to deal with it in an exhaustive manner. References are provided by way of illustration of past trends and emerging issues to be explored.

Rethinking appropriate categories to capture shifting boundaries

The process of liberalisation has contributed to the blurring or shifting of boundaries in various areas such as, for example, boundaries which existed between public as opposed to private spheres of authority and influence, as a result of the transferring of what were previously considered public functions (rulemaking, roads, service delivery, security, etc.) to private actors and notably large transnational mining companies. Similarly, the demarcation between domestic as opposed to international spheres of activity has been obscured, rendering difficult the capacity to trace revenue flows which are transferred to offshore accounts.[10]

The stark dichotomy introduced concerning artisanal and small-scale mining (ASM)[11] as opposed to industrial or large-scale mining (LSM), and the manner in which these categories have been used by both policy-makers and researchers may be seen in large part to reflect the development agenda and strategies which have sought to open up the African mining sector to foreign investment. The implications of the strategies which have been introduced and the paradoxes which they have produced will serve as the starting point to discussion. The Ghanaian experience is particularly interesting in this regard. According to Hilson and Okoh:

> [...] despite attracting US$11.5 billion in investment in the sector between 1983 and 2011, and producing a record 3.6 million ounces of gold in 2011, Ghana's large-scale mining 'enclaves' only employ, collectively, 28,000 people (approximately one job per US$400,000 invested) and have few links to the local economy (Aryee, 2012). By comparison, ASM employs over one million people directly in the country and millions more in downstream industries, and contributes close to 30 per cent of national gold output.[12]

In view of this paradoxical situation and in order to explain the lack of emphasis on the potential of ASM in terms of employment possibilities and improving livelihoods, the authors point to the 'prioritisation' of industrial mining introduced with the liberalisation of the sector. Hilson and McQuilken (2014) make a similar suggestion[13]. As a result, ASM is perceived:

> [...] as a nuisance and impediment to broader plans to develop large-scale projects. The focus has rather been on the sector's negative aspects, including its abysmal health and safety record and significant environmental footprint, but with little acknowledgement that these problems are in large part a result of the sector's largely-informal character.[14]

Reforms conducted with the aim of liberalising the mining sector at the benefit of large-scale activities have also contributed to the increase of unregulated artisanal mining. Banchirigah has pointed out:

> Several factors have contributed to the recent increase of artisanal mining activity in sub-Saharan Africa, many of which are linked to reforms. These include the unavailability of land for acquisition by small-scale miners, a result of large concessions being leased out to large-scale mining companies; the absence of transferable mineral rights, which prevents the acquisition of sections of concessions awarded to large-scale mining companies that prove unfeasible to work but are at the same time, suitable for small-scale mining; and attempts to formalise small-scale mining, which provide little incentive to operate within the legal domain, and therefore push more people into illegal, artisanal activity[15].

As opposed to the persistence of the largely unregulated nature of ASM activities, which may be seen as a central cause for the severe problems which have accompanied them, operations in the industrial sector were, according to World Bank 1992 recommendations, to be attracted thanks to the introduction of clear rules and guarantees regarding exploration and mining rights, majority ownership for the private investor, taxation, foreign exchanges, escrow accounts, and profit and capital repatriation (that these rules were to be set out before exploration takes place). Mining codes were also to "assure companies that they will have the right to mine following successful exploration, and that they will be permitted to transfer or trade exploration or mining licenses subject to explicit criteria".[16] In terms of regulatory guarantees, the contrast could hardly be starker. The industrial sector was set to develop not only as distinct from ASM, but in fact is in opposition to it. The severity of the problems resulting from such strategies did not take long to become apparent. Their persistence may be attributed to the absence of redefinition of policies in this area which in turn reflected the constellation of structural relations of power which have kept in place the particular hierarchy of norms outlined above.

The point we wish to underline here is that the dichotomy introduced by the reforms, enshrined in the regulatory frameworks and promoted by the strategies in the sector, have created a distinction which while very present in the representations we have of the sector, is in fact somewhat removed from empirical reality. The following description of activities at Obuasi brings this out clearly where:

> [...] perhaps more so than anywhere else in Ghana, illegal artisanal miners work directly alongside the large-scale mining company, in this case AngloGold Ashanti, for the most part, undisturbed, due to arrangements such as that between management and groups such as the Obuasi Municipal Artisanal Miners Association. Most of the town's businesses, including shops, commercial vehicle hire, hotels and petrol stations are owned, financed or run by miners (both people working at AngloGold or engaged in ASM). Although there are some differences in lifestyles between the people (Ghanaians) working at the company and those engaged in *galamsey,* [[17]] this distinction is often blurred, largely because many now involved in the latter at one point worked at the mine. A local assembly member reflected on why in an interview, reinforcing points raised earlier about how enclaves such as large-scale gold mines constrain development:
>
> 'The company used to have about 13,000 to 15,000 workers, but now the company has reduced its workforce to about 4,000, so you can imagine what happens to the laid off workers. How do they find their livelihood? Most of them are still living within the communities. These laid-off workers happened to be underground workers who know the geology of the area, so they have resorted to *galamsey* or partnered with the *galamsey* guys [...]. It is therefore no surprise that one cannot distinguish between the former workers of AngloGold Ashanti now engaged in *galamsey* and the present workers of the company [...]. We are living in a community where mining is the ultimate job. Even teachers and civil servants are not satisfied with their income and would prefer to be in the mines'. (Interview, local assembly member, Obuasi, 14 April 2011)[18]

A growing number of authors have questioned the relevance of the dichotomy between ASM and LSM. For example, in his analysis of the mining sector in Philippines, Verbrugge concluded that a dualist perspective of LSM and ASM hides the fact that small-scale mining operations are heterogeneous and that their interactions with large-scale mining operations are very complex and multiple[19]. He also points out that small-scale mining expansion in Philippines was partly due to the crisis of the industrial sector since many mineworkers turned towards ASM activities while large-scale mines experienced a downturn in the 1980s[20].

The 'prioritisation' of industrial mining has clearly led to a situation of policy neglect – a point to which we shall return in the third section below. The above analysis on Ghana, along with the work of authors such as Verbrugge, lead us to suggest that not only from the standpoint of contributing to a more accurate analysis of employment patterns and livelihood strategies, but also from the perspective of ensuring a more comprehensive and exact understanding of the complex links among various forms of mining activities and livelihood strategies more generally[21], a nuanced and historically grounded approach to the articulation of activities within the mining sector as a whole needs to be adopted. The ongoing renewal of approaches has contributed to addressing another potential shortcoming of research which seeks to contribute to ASM-policy formulation, the failure to meaningfully differentiate *within* the ASM-sector, both in terms of actors and practices. In this regard, an increasing body of literature has drawn attention to the wide variance in types of roles and livelihoods, of the structures of mining communities and value chains.[22] These concerns suggest the need for careful scrutiny of the categories and boundaries which have been introduced and underline the fact that their analysis cannot be undertaken without a deep understanding of the historical processes of which they are a part and the specific social, cultural and political contexts in which ASM activities take place.[23]

Revisiting approaches

As illustrated by the examples in the previous section, the revisiting of the analytical categories used and the boundaries between them depends in turn on the renewal of approaches. Clearly, these need to be holistic and able to take account of the specificities of different countries which are characterised by very diverse policy traditions, trends and objectives, reflecting an enormous variance with regard to history and context. Several academic contributions have underlined the need to take into account the heterogeneity of the contexts in which ASM takes place and to redefine theoretical frameworks[24]. For example, Siegel and Veiga who focus specifically on Uganda have argued that formalising ASM should be understood in the context of Hernando De Soto's theory of 'extralegality'. In this framework, formalisation is the means of absorbing existing customary practices – developed informally by miners – into the mainstream of a country's legal and economic affairs.[25] While it is not possible in this short article to develop additional examples, it is nonetheless important to underline the diversity of existing approaches, notably with regard to anthropology-oriented authors on ASM issues.[26] Recent research in Senegal will be developed a little more thoroughly not only because it illustrates the ongoing redefining of approaches which have been applied to the study of artisanal and small-scale mining so that these become more multifaceted and participative and in this case, includes legal dimensions, but also because it is situated in a country where interesting policy reforms are taking place. (See third section).

In the conclusion to his book *Des pierres dans les mortiers et non du maïs* [Stones in the Mortars Rather Than Corn] (our translation), Bakary Doucouré calls for the adoption of a holistic socio-anthropological approach which takes account of economic, professional, political and environmental dimensions. He underlines that as methodological approaches in the social sciences are not exclusive, what is important will be the articulation of approaches from various disciplines which draw on methodologies which may be qualitative, quantitative and participative.[27] In his perspective, there are different points of entry to this area of research. They are complementary and their articulation will require further research.

There is a wide variety of disciplinary perspectives which one could adopt to study the developmental implications of activities in this area. Our own work draws on contributions from two currents - legal pluralism and heterodox political economy (IPE). With regard to legal pluralism this approach allows us to take account of multiple, overlapping and at times conflicting sets of jurisdictions which characterise the mining sector, particularly, for historical reasons, in the countries of Africa and Latin America.[28] In this regard, several analysts have already stressed the need to recognise the regulatory and legal complexities as well as the multiple actors involved in ASM [29] and hence the potential usefulness of legal pluralism.

From heterodox IPE we have found valuable the contributions of writers such as Strange[30] and Chavagneux[31] and notably notions such as *structural power*. For example, in our examination of the reform of regulatory frameworks, we find useful the notion of structural relations of power which refers to the capability to create essential rules, norms and modes of operation for various dimensions of a particular system whether national or international. The analysis of power and the distinction between relational and structural power in particular was at the centre of most of Susan Strange's later theoretical writings about the global political economy.[32]

Whether with regard to mining activities in specific African countries or concerning specific mining negotiations in Canada, our research has allowed us to see mining regimes not only as regulatory frameworks but above all as a key component of the structure of power that conditions relations among the actors involved in the sector. Questioning the relevance of the prevailing paradigm which prioritises the implementation of property rights regimes in artisanal and small-scale mining, Spiegel raised concerns about the structure of power in the Indonesian mining sector:

> Civil society organizations and government agencies should pursue development planning in ways that do not marginalize poorer workers in the aforementioned 'perverse' and 'delusional' ways by championing property rights systems that privilege powerful elites at the expense of local rights claims. Empowering district institutions with greater capacities to regulate and study the mining sector, with clear mandates for assistance and monitoring, should be seen as vital to ensure that mineworkers participate in guiding decision-making processes that affect them[33].

It is therefore in this historical perspective which takes account of the origins and the design of regulatory frameworks, the resulting institutional arrangements and the relations of power and authority which shape such frameworks, that I propose to set the broader discussion of 'governance', the improvement of which is often put forward as a solution to the complex issues facing the mining sector as a whole.

For many, the notion of governance is in fact quite problematic because as used at present by a wide range of actors, it most often leads to a certain technicisation and de-politisation of social and political processes. In order to distance from such a narrow notion of 'governance' which reflects more procedural, managerial and technical preoccupations, different contributions exist which attempt to provide a more encompassing and satisfactory notion of governance of mining activities. For example, alternative approaches have drawn attention to the recognition that there are different arenas of governance which may be captured by the notion of 'governance flows'. As Bebbington has pointed out:

> While extractive industry is governed in many spaces, Bridge's (2008) exploration of the points of contact between Global Production Networks and regional approaches to resources and development suggests that one useful distinction might be that between the governance of flows and the governance of territories. While the most obvious flows in extractive industries are of capital (investment and profit) and commodities, flows of materials (e.g. carbon), information, non-profit resources and solidarities are also central elements of the political economy of extraction.[34]

Speaking of the flows and territory, the same author goes on to suggest:

> These two domains of governance are in practice deeply intertwined. The governance of territories influences flows, just as the governance of flows conditions how territories are governed. This perspective adds richness and depth to the study of articulations and resistances among governance initiatives operating at different scales, and the formation of governance assemblages linking both territories and flows.[35]

An important area for future research will clearly be to analyse the different forms of articulation which emerge between these different arenas of governance and unravel the ways in which these arenas interact with one another. While governance flows in the industrial sector take place largely through private agreements, most often contracts between states and companies, mining regimes, in principle, articulate the conditions under which all mining operations, including in the artisanal and small-scale sector, take place over national space and beyond. Institutional arrangements are obviously a key intersection for understanding the articulation between different arenas of governance such as these and consequently point to the usefulness of a more encompassing approach.

A possible avenue in this regard is through the adoption of the notion of 'modes of governance' and its use in the analysis of the sector as a whole identified in particular historical settings. This notion proved very useful in our recent volume[36] as it provided an analytical framework for the series of case studies. Such a perspective not only allows taking account of particular sets of institutional arrangements, regulatory frameworks and hence property rights, but as a result, carries the possibility of contributing to a better understanding of the conditions under which these may eventually be transformed through political processes involving local actors.

Hence, and in opposition to a perspective of analysing 'governance' as 'the management of the social order', and in the case of the topic of this collection of articles, the governance of mineral resources as essentially the management of the resource a more holistic approach is provided by the notion of 'modes of governance'. This broader notion entails taking account of the particular forms of regulation (the mining regime, the fiscal regime, the environmental legislation, the labour legislation and so on) that determine the conditions of exploitation of mineral resources for any particular project, region or country, without specifying a particular sub-sectors whether artisanal, small scale or industrial. The definition puts emphasis moreover on the actors responsible for the forms of regulation where these exist, whether internal or external, public or private. In so doing, the notion permits identifying the specific social relations which are at the origin of institutional arrangements, as well as the role of the actors concerned in the decision-making processes which emanate from these arrangements. In a nutshell, the mode of governance in the mining sector represents the sum of the forms of regulation for each of the related dimensions (economic, social, political and environmental), which determines, in any given period, the conditions of exploitation of mining resources.[37] This notion should not be seen as a rigid 'blue print' but rather a tool which informs a framework of analysis in which the interaction of forms of regulation, institutional arrangements, relations of power and influence are seen as key components shaping a common approach. By drawing attention among areas to the importance of regulatory frameworks, this approach permits examining the processes which give rise to such frameworks, the norms on which they are based, where these norms have come from, how they are implemented and through what policy processes they might be changed. This observation meshes well with the methodological conclusion of Doucouré[38] who draws attention to the central importance of the juridical situation of artisanal mining, whether legal or not. This analyst recommends taking as a starting point the analytical framework provided by a socio-anthropological approach, with which must be articulated legal information, whether concerning articles of law, regulations, decrees, codes, laws and contracts. This would permit the adoption of research strategies which might be qualified as "*socio-anthropologie juridique*" – socio-anthropological legal studies.[39]

As revealed by the above, the legal dimensions of artisanal and small-scale mining are at the very heart of discussions and debates concerning regulatory and institutional reforms and consequently of the renewal of policies in this area. By placing an understanding of regulatory frameworks (the mining regime, the fiscal regime, the environmental legislation, the labour legislation, etc.) at the centre of analysis, the notion of 'modes of governance' may be seen as a particularly suitable point of entry for revisiting the role of ASM in past livelihood and development strategies. For such an approach implies understanding the relations of

Table 1. African countries with more than 100,000 ASM operators.

Country	ASM	Estimated dependants
Angola	150,000	900,000
Burkina Faso	200,000	1,000,000
Central African Republic	400,000	2,400,000
Chad	100,000	600,000
Côte d'Ivoire	100,000	600,000
Democratic Republic of the Congo	200,000	1,200,000
Eritrea	400,000	2,400,000
Ethiopia	500,000	3,000,000
Ghana	1,100,000	4,400,000
Guinea	300,000	1,500,000
Liberia	100,000	600,000
Madagascar	500,000	2,500,000
Mali	400,000	2,400,000
Mozambique	100,000	1,200,000
Niger	450,000	2,700,000
Nigeria	500,000	2,500,000
Sierra Leone	300,000	1,800,000
Sudan	200,000	1,200,000
Tanzania	1,500,000	9,000,000
Uganda	150,000	900,000
Zimbabwe	500,000	3 000,000

Sources: Estimates based on CASM, ASM statistics for Africa. (Reproduced from Table 5.1 in the Minerals and Africa's Development Report. UNECA and AU, *Minerals and Africa's Development*, 69.)

influence and power, political processes and institutional arrangements which explain the current heritage, the particular problems which have emerged and consequently, the conditions necessary for renewing policies in this area of activities.

Renewing institutional reform and policy processes from a country-specific and participative perspective

Perhaps the most striking feature concerning ASM is the degree to which mining activities of this nature which in Africa alone concern not just hundreds of thousands but millions of workers and their families[40] (see Table 1), have until recently continued to operate most often in the absence of an effective regulatory frameworks.[41]

The fact that most governments now have in place regulations for ASM, i.e. the activity is legal through the purchase of a licence and there is usually a short section in the mineral policy, does not in itself translate into actually supporting and incentivising miners to formalise and enter the legal domain. While possible reasons for this have been suggested above and concern the norms introduced with measures of liberalisation which prioritised and even opposed industrialised mining to ASM, the heritage of this situation is worth illustrating and will be briefly summarised by citations from the UNECA 2011 Report[42], before several avenues of policy reform are examined. It should be noted that this report was commissioned to provide concrete recommendations to permit the implementation of the African Mining Vision adopted by the Heads of States of the African Union in 2009. Among its recommendations, it called for:

> A mining sector that harnesses the potential of artisanal and small-scale mining to stimulate local/national entrepreneurship, improve livelihoods and advance integrated rural social and economic development.[43]

By way of illustration of the lack of formal regulatory frameworks, research on the Democratic Republic of the Congo (DRC) conducted in the Southern Kivu and published in 2009:

> [...] began with the discovery that the artisanal extraction of minerals, despite its importance to entire regions of the DRC, is hardly taken into account by the 2002 Mining Code. Indeed, the Code only devotes a few articles to it (art. 109 to 128), or 20 articles out of a total 344, which cover the whole of the industry from the mining site to the mineral trading house (export stage). This is hardly surprising given that the Mining Code was first drafted with a view to attracting investors to industrially develop the immense mineral resources that state companies such as Gécamines[1] or SODIMICO[2] were unable to develop on their own due to a lack of financial resources. Similarly, while it is now argued that artisanal miners should organize and improve their operations to raise the technological and legal level of small-scale mining, the Mining Code only briefly addresses the status of small-scale mining, devoting just 12 articles to it (art. 97 to 108).[44]

The study came to the conclusion which was demonstrated time and time again not only in the DRC but elsewhere in many other African countries where ASM is practiced: 'that there are major discrepancies between realities in the field and existing artisanal mining norms'.[45] In this regard, Teschner has pointed out in his study on the Ghanaian mining sector, that the dichotomical perspective on formal and informal mining no longer reflects the reality of mining activities. Registered and unregistered small-mining operations are both engaged in informal activities (use of explosives, informal employment, informal financial sources, etc.). Rather than using a dichotomical analysis which aims to differentiate informal and formal mining operations, this author suggests that it is more appropriate to see the ASM sector as a highly intertwined group of 'semi-formal' sectors operating with varying degrees of legal registrations as well as technological and market innovations of varying legal status.[46]

The divergence observed in countless situations concerning this area of activity is summed up by the 'a-legal' neologism, a concept coined by Nicholas Garrett and Harrison Mitchell based on the Greek privative alpha prefix to describe a situation that is only illegal because the state does not create the necessary conditions for actors to comply with the law.[47]

While the problems confronting those involved in ASM are severe and many and include limited technical capacity and access to appropriate technology (and consequent environmental degradation); lack of finance; inadequate access to exploration and mining areas; difficulties in accessing the market; issues associated with conflict minerals; and women and child labour concerns, rather than addressing specific issues such as these which are the object of a very rich and evolving body of literature[48], emphasis here will be placed on the inadequacy of policy and regulatory frameworks to respond to such challenges more generally. The observations and recommendations of the 2011 UNECA report merit attention:

> The absence of specific policy frameworks facilitating the emergence or growth of sustainable ASM operations is a major constraint in most African countries. In some countries, small-scale mining policy and regulations fall under general mining policy, which makes no distinction between LSM (large-scale mining) and ASM. Thus the peculiar challenges in small-scale mining do not receive the attention that they deserve. Even in countries with a separate small-scale mining policy, the procedures for acquiring licences are generally cumbersome, which becomes a barrier to formalization. Without specific frameworks, ASM operators face challenges in getting mining rights.
>
> Even when operators have formal access, the mining rights rarely provide for security of tenure: their duration is short and the size of the licence area usually small.[49]

With regard more specifically to the links between land governance and property rights[50] and the difficulties which arise concerning access to exploration and mining areas, the UNECA Report notes:

> ASM policies in many countries provide for designation of land or areas for exploration and mining activities. If properly implemented, such policies have the potential to reduce tensions between ASM and LSM. Such designation might result in fewer conflicts over exploration and mining areas and could create space for beneficial collaboration by the two sectors. Among the many benefits could be a reduction in environmental degradation associated with haphazard mining.
>
> Yet such land designation has so far failed to solve these problems, largely because few of the demarcated areas have been properly surveyed for ASM. Detailed geological information to prove their suitability for ASM is rare, weakening the relevance of "reservation policies" for ASM. The challenge for policymakers is to find ways to conduct or fund comprehensive exploration to determine suitable areas for ASM and then provide such information to the miners.[51]

While very useful in that this overview presents a comprehensive mapping of the different problems facing ASM, the focus on specific questions such as land designation illustrates a perspective quite different from one that adopts the notion of "modes of governance" which would entail resituating the analysis of issues such as land governance within the broader context of the regulatory frameworks and historical settings which are specific to particular countries or regions. Moreover, although there have been numerous recommendations in response to these challenges, as for example those of the Yaoundé Vision on ASM adopted in 2002,[52] of the UNECA in 2009[53] or of the donor community such as the initiative Communities and Small-Scale Mining (CASM),[54] it is striking to see on the one hand, the degree to which these approaches have remained disaggregated rather than comprehensive, and on the other, the extent to which little overall progress has been achieved in the introduction of reforms in this area. For example, the Compendium on Best Practice for ASM published by UNECA[55] provided a catalogue of experiences from Africa. At the national level, certain countries 'have mainstreamed ASM into poverty reduction strategies. Some have passed legislation to simplify legal requirements, and others, such as Ghana and Zambia, have changed the law to improve the environment for ASM'.[56] As has been well documented however, even in countries such as Ghana, the complexities facing artisanal miners in order to obtain mining rights, the duration of mining rights and size of concessions, remain very important constraints.[57] More generally, in terms of the shaping of policy, to date, the approaches adopted by most countries continue to be directed to specific challenges rather than being comprehensive.

In response to this heritage there has emerged a call for new responses including the 'formalisation' of the sector. In order for such measures to be in a position to address the depth and complexity of issues at hand, such a renewal implies the adoption of more holistic approaches, informed by more comprehensive information, based on country specific situations, gathered in a perspective which as suggested above, could draw on socio-anthropological legal studies. Examples of ongoing research in this area include, among others, the work of Van Bockstael on Liberia, Siegel and Veiga on Uganda and the extralegal economy theory, Verbrugge on Southern Philippines, or again Banchirigah on Ghana. In the end however, a renewal of policy responses also entails the political will and space to change the prioritisation which in the past has disregarded and marginalised this area of activity and policy.

With regard to a potential new avenue, while there are ongoing discussions concerning the possibility of productive collaboration between ASM and LSM, as the UNECA 2011 report pointed out, 'important policy and structural issues need to be resolved before such mentorship could deliver its full potential. The approach needs to be adapted for artisanal miners, who are not as organized as small-scale miners'.[58]

It is clear that policy solutions such as these will depend on the prior introduction of appropriate regulatory frameworks and the capacity to monitor, adjust and enforce such frameworks. To this end, the emphasis placed on regulatory frameworks by approaches which privilege a more holistic perspective, such as that proposed by the notion of modes of governance, would seem a useful starting point. It is in fact that which appears to inform the recommendations of the UNECA which underline that ASM policy has moreover to be formulated and implemented as part of a broad rural development strategy. The 2011 Report proposes numerous concrete recommendations of which the first and overarching one is 'Regularizing informal ASM', after which follow: 'Simplifying and decentralizing procedures for acquiring ASM rights; Providing a realistic implementation plan, including institutional capacity enhancement'.[59]

Among current initiatives to renew policies in the area of ASM at a national level, that of Senegal illustrates several important aspects brought out in this article, notably with regard to the need to adopt a broader developmental approach. Spurred by the current Director of the Ministry of Mines and Geology, the reorganisation of ASM which began in August 2012 has until now followed three stages. The first phase was one of consultation. Field trips were organised in order to analyse the problems and the issues to be addressed. Furthermore, meetings were held with all of the ministries concerned (mines, health, security, commerce, the environment, etc.), as well as with the mining companies operating in areas which are the sites of artisanal mining. Information and awareness-raising sessions were organised with artisanal miners in order to hear of their concerns. This phase ended with the holding of an inter-ministerial Council meeting on 22 November 2012 where a series of measures to be taken were set out.

The second phase was that of implementation. It was ushered in by a ministerial decree regulating artisanal mining and introducing a card or permit for artisanal miners. The regulatory framework was completed on the one hand, by a decree which identified special zones which will be reserved for artisanal activities and on the other, by the adoption of procedures to regulate the sale of precious metals derived from ASM. The third phase will be one of the accompanying of the artisanal sector through the introduction of an appropriate institutional framework and measures of support and encouragement of artisanal mining in such a manner that these activities become viable and sustainable.

While obviously too early to draw conclusions concerning the Senegalese initiative, it is interesting on several accounts. First is the fact that it comes as a result of the recognition that a new priority must be given to the redefining of the regulatory framework in this area. Second is the fact that it is driven by specialists who have an intimate knowledge of the mining sector as a whole and ASM in particular.[60] Third, it is not the project of merely one ministry but an initiative launched that has an 'all of government approach' and the endorsement of a wide range of ministries in order to deal with the multifaceted nature of the issues raised. Also, in the first consolation phase, in order to minimise their conflicts with artisanal miners and in exchange for a measure that would minimise risk to their operations, industrial companies were persuaded to allocate part of their concessions to artisanal miners. Furthermore, in exchange for access to secured land, the proposal involved persuading artisanal miners that they would have to obtain a permit to operate. This was done by providing public financial support to miners in order that they were in a position to obtain the necessary identification cards required as a precondition to be granted a permit. Other aspects of the reorganization, which is currently ongoing, entail the setting up of buyers (*comptoirs*) which will pay the taxes which would normally be paid by the miners, a procedure which frees the latter from paying these levies and means that the buyers have an interest in ensuring that the miners from whom they purchase ore are registered and have a permit. Finally, and as elsewhere in Africa, as captured by the title of the book by Doucouré, in a country where the struggle to ensure a means of livelihood has created competition for access to land between artisanal mining and agriculture, the new regulatory framework provides that artisanal mining is to be an off-season activity and not to be carried out in the rainy season.

Conclusion

Although still at an early stage of its implementation, the Senegalese initiative suggests interesting responses to several of the points raised in this article. First is the importance of rethinking boundaries. In the example given, this has been done by conceiving a new regulatory framework for ASM which links it to LSM by ensuring access to land and mining rights for the activities of the former, while minimising risks for the operations of the latter, in a manner which articulates rather than opposes activities within the sector as a whole. It illustrates also a redefining of approaches both because it represents an 'all of government' initiative and because it adopts a holistic approach which involves addressing concerns raised by a wide range of areas including health, the environment, access to land title and the commercialisation of output. Consideration of these interconnected dimensions can be seen as the basis and an important precondition for the renewal of policies in this area. Perhaps above all, the active intervention of the government to regulate the sector by requiring concessions from industrial mining companies, signals a revisiting of norms which have so far given overriding priority to large-scale mines. Without wanting to draw overly hasty conclusions, it appears safe to say that the intervention of the state in this area to redefine norms which have predominated in the past, may be seen as a step in a process leading to a new demarcation of former spheres of authority and influence and potentially, the beginning of the reshaping of structural relations enshrined in regulatory frameworks which to date have been very highly skewed to the detriment of other than industrial forms of activity in the mining sector. Whether the measures of the 'legal integration' of ASM taken so far by Senegal will be able to address the considerations raised by Fischer which have the potential to impact on social exclusion is still to be seen. Such factors include, among others, the capacity of reforms to take account of local institutional and organizational formations, existing inequalities within the ASM sector, social differentiation and power disparities.[61]

While still preliminary and requiring further work, this article has argued that notions such as modes of governance and structural relations of power may represent useful tools in order to explore the interconnections among different facets of ongoing debates. These include the past dualist understanding of the mining sector, inherited legal complexities, the need for richer approaches to governance and the critical importance of integrating ASM-policies in broader rural development strategies. Because of their holistic nature, these concepts open potential avenues for deepening understanding of certain conceptual and methodological obstacles to reformulating policy proposal from a broader developmental perspective, notably with regard to the links between land governance, property rights and mining. Their potential rests among other things in the fact that they oblige us to begin with a careful examination of the heritage, the regulatory frameworks and policies responsible for very boundaries and categories which we use and which have led us to distinguish artisanal, small-scale and industrial operations and above all, to question the viability of such categories and past strategies for the future.

Disclosure statement

No potential conflict of interest was reported by the author.

Notes

1. World Bank, *Strategy for African Mining*, 16.
2. Ibid.
3. Naito, K. et al., *Review of Legal and Fiscal Frameworks*.
4. World Bank, *Strategy for African Mining*, 27.
5. Ibid., 54.
6. Campbell, *Mining in Africa*; Campbell, *Regulating Mining in Africa*.
7. Campbell, "CSR and Development in Africa."
8. Campbell, "Revisiting the Reform Process."
9. Gagné-Ouellet, "Regulatory Framework Review," 52.
10. Campbell, *Modes of Governance*.
11. According to the UNECA 'There is no consensus on what constitutes a small-scale mining operation; neither is the boundary between ASM operations clearly defined. This is partly because definitions vary by country'. UNECA and AU. *Minerals and Africa's Development*, 67. According to Hilson and Okoh: "Compared to their capital-intensive large-scale counterparts, ASM operations are rudimentary in design but provide far more jobs; target mainly alluvial and/or marginal near-surface deposits which have little appeal to foreign companies; and are propelled, almost exclusively, by finances sourced from the informal economy." Hilson and Okoh, "Artisanal Mining in Ghana," 139; quoting ILO, *Social and Labour Issues*; and Hentschel, Hruschka, and Priester, "Global Report on Artisanal."
12. Hilson and Okoh, "Artisanal Mining in Ghana," 144.
13. Hilson and McQuilken, "Four Decades of Support."
14. Hilson and Okoh, "Artisanal Mining in Ghana," 140.
15. Banchirigah, "How Have Reforms Fuelled."

16. Hilson and Okoh, "Artisanal Mining in Ghana," 141.
17. "A label used in Ghana for illegal mining, the literal meaning of which is 'gather them and sell.'" Ibid., 160.
18. Ibid., 156, 157.
19. Verbrugge, "The Economic Logic."
20. Verbrugge and Besmanos, *Undermining the Myths*.
21. The very long time it took for policy-makers to recognise the links between ASM, livelihoods and development has become a recurrent theme in the literature: ILO, *Social and labour issues*.
22. Among the authors who have contributed to this ongoing renewal of approaches: Fisher, "Occupying the Margins"; Hilson, "Artisanal mining, smallholder farming"; Hilson and Van Bockstael, "Diamond mining, rice farming"; Maconachie, "Re-agrarianizing livelihoods"; Maconachie and Binns, "Farming Miners"; Yakovleva, "Perspectives on female participation"; and Levin and Gberie, "Dealing for Development".
23. On the heterogeneity of ASM operations and the influence or social, cultural and political contexts: Fisher, "Occupying the Margins"; or again on the intertwining of farming and mining and importance of understanding the local dynamics: Maconachie and Binns, "Farming Miners" or "mining farmers?".
24. Clausen, Barretto and Attaran, "Property Rights Theory"; Siegel and Veiga, "Artisanal and Small-scale Mining"; Banchirigah, "Challenges with Eradicating Illegal Mining"; Spiegel. "Governance Institutions"; Fisher, "Occupying the Margins"; Hinton, "Communities and Small-scale Mining"; and Hilson and Pardie, "Mercury: An Agent of Poverty."
25. Siegel and Veiga, "Artisanal and Small-scale Mining."
26. A few examples of the diversity of anthropology-oriented authors on ASM issues include D'Angelo, "Who owns the diamonds?"; De Theije and Heemskerk, "Moving frontiers in the Amazon"; Fisher, "Occupying the Margins"; and Luning, "Liberalisation Gold Mining Sector."
27. Doucouré, *Des pierres dans les mortiers*, 147.
28. Szablowski, *Transnational Law and Local Struggles*.
29. Fisher, "Occupying the Margins"; Spiegel, "Governance Institutions"; Teschner, "Small-scale Mining in Ghana"; and Verbrugge, "The Economic Logic."
30. Strange, *States and Markets*; Strange, *The Retreat of the State*; and Strange, "International Political Economy."
31. Chavagneux, "Peut-on maîtriser la mondialisation?".
32. Strange, *States and Markets*, 24, 25.
33. Spiegel, "Governance Institutions".
34. Bebbington, *Natural Resource Extraction*.
35. Ibid.
36. Campbell, *Modes of Governance*.
37. Belem, "Quelle gouvernance".
38. Doucouré, *Des pierres dans les mortiers*.
39. Ibid, 34 (Our translation).
40. According to the International Council on Mining and Metals (ICMM): 'ASM is a form of subsistence mining providing a direct or indirect livelihood for approximately 100 million people in the developing world. It often takes place alongside large-scale operations'. ICMM, "Artisanal and Small-scale Mining". According to Hilson and Okoh: 'ASM in sub-Saharan Africa, employs at least 10 million people directly and creates jobs for many millions more in downstream industries'. Hilson and Okoh, "Artisanal Mining in Ghana," 139.
41. Perks, "The Inclusion of Artisanal."
42. UNECA and AU, *Minerals and Africa's Development*.
43. AU, *African Mining Vision*.
44. De Failly, Ntakobajira, and Shonja, "Traceability of Local Mining Flows", 164. Notes in the citation:

    Générale des Carrières et des Mines, a state owned company created in 1967 as a successor to the very famous Union Minière du Haut-Katanga, a private corporation that stemmed from the Société Générale de Belgique holding company.

Société de Développement Industriel et Minier au Congo, created in 1967 and purchased by Gécamines in 1987.

45. Ibid., 164. See also Ndimubanzi Ngoroba, "Décalage entre les textes légaux".
46. Teschner, "Small-scale Mining in Ghana."
47. De Failly, Ntakobajira, and Shonja, "Traceability of Local Mining Flows," 165.
48. UNECA and AU, *Minerals and Africa's Development*, 70; Yakovleva, "Perspectives on female participation"; African Minerals Development Centre, *African Women in Artisanal*; Hinton, Veiga, and Beinhoff, "Women and Artisanal Mining"; Jennings, "Addressing Labour and Social Issues"; Jønsson, Appel and Chibunda, "A Matter of Approach"; Clifford, "Potential Repercussions of a Mercury Ban"; Bose-O'Reilly et al., "Mercury in Breast Milk"; and Hilson and Ackah-Baidoo, "Can Microcredit Services Alleviate Hardship."
49. De Failly, Ntakobajira, and Shonja, "Traceability of Local Mining Flows," 165.
50. While beyond the scope of this paper, an important aspect of this subject concerns the particular ways in which existing practices in this area disadvantage women active in the sector in important ways. UNECA and AU. *Minerals and Africa's Development*, 74. The same source also suggests that: "From a regulatory viewpoint, the provision of women friendly facilities and technology in mining areas could be made mandatory and a legal requirement for issuing a mining permit or allocating mining rights." Ibid., 78, 79.
51. Ibid., 72.
52. AU, *African Mining Vision*, 26–29.
53. UNECA, "Viable SME mining."
54. 'Communities and Small-scale Mining (CASM) is chaired by the UK government's Department for International Development and is housed at the World Bank in Washington D.C. It was launched in March 2001 in response to international recognition of the need for an integrated approach to address the challenges facing ASM communities and for improved coordination between institutions funding and executing assistance. CASM began as a multi-donor networking and coordination facility that would engage with practicing miners, their associations and communities, governments and non-governmental organizations, and development assistance agencies'. CASM, "Information."
55. UNECA, "Compendium on Best Practices."
56. UNECA and AU, *Minerals and Africa's Development*, 75.
57. Hilson and Okoh, "Artisanal Mining in Ghana."
58. UNECA and AU, *Minerals and Africa's Development*, 78.
59. Ibid., 79; McQuilken and Hilson, "Artisanal and Small-scale Mining in Ghana."
60. Before becoming the first Director of the Ministry of Mines and Geology and the person who heads up the current reorganisation of ASM in Senegal, Ousmane Cissé was between 2000 and 2002, the geologist in charge of the monitoring and development of ASM and, from 2002 to 2007, acted as the Director of the Regional Office of Mines and Geology for Thiès (Chef du service régional des mines et de la géologie de Thiès). Subsequently, he was the counterpart to the head of the project responsible for the setting up of the Documentation Centre and the Mines Registry Office [Centre de Documentation et du Cadastre Minier (CDCM) of the Direction des Mines et de la Géologie].
61. Fisher, "Occupying the Margins", 756.

Bibliography

AMDC (African Minerals Development Centre). *African Women in Artisanal and Small-scale Mining. Special Report by the AMDC*. Addis Ababa: African Minerals Development Centre, African Union, 2015.

AU (African Union). *African Mining Vision*. Addis Ababa: African Union and United Nations Economic Commission for Africa, 2009.

Aryee, B. N. A. "Contribution of the Minerals and Mining Sector to National Development: Ghana's Experiment." *GREAT Insights* 1, no. 5 (2012). http://ecdpm.org/great-insights/extractive-sector-for-development/contribution-minerals-mining-sector-national-development-ghanas-experiment/.

Banchirigah, S. M. "Challenges with Eradicating Illegal Mining in Ghana: A Perspective from the Grassroots." *Resources Policy* 33 (2008): 29–38.

Banchirigah, S. M. "How Have Reforms Fuelled the Expansion of Artisanal Mining? Evidence from Sub-Saharan Africa" *Resources Policy* 31 (2006): 165–171.

Bebbington, A. *Natural Resource Extraction and the Possibilities of Inclusive Development: Politics across Space and Time*. ESID Working Paper 21. Manchester: Effective States and Inclusive Development, 2013.

Belem, G. "Quelle gouvernance pour la mise en œuvre du développement durable? L'expérience de l'industrie minière du Mali [What Form of Governance Can Ensure Sustainable Development? The Experience of the Malian Mining Industry]." PhD Thesis in Environmental Sciences, Université du Québec à Montréal, 2009.

Bose-O'Reilly, S., B. Lettmeier, G. Roider, U. Siebert, and G. Drasch. "Mercury in Breast Milk – A Health Hazard for Infants in Gold Mining Areas?" *International Journal of Hygiene and Environmental Health*, 211, no. 5–6 (2008): 615–623.

Bridge, G. "Global Production Networks and the Extractive Sector: Governing Resource-Based Development." *Journal of Economic Geography* 8 (2008): 389–419.

Campbell, B., ed. *Mining in Africa: Regulation and Development*. London/Ottawa/Uppsala: Pluto Press/International Development Research Center/The Nordic Africa Institute, 2009. http://www.idrc.ca/EN/Resources/Publications/openebooks/450-5/index.html.

Campbell, B., ed. *Modes of Governance and Revenue Flows in African Mining*, International Political Economy Series. Basingstoke: Palgrave Macmillan, 2013.

Campbell, B., ed. *Regulating Mining in Africa: For Whose Benefit?* Discussion Paper 26. Uppsala: Nordiska Afrikainstitutet, 2004.

Campbell, B. "CSR and Development in Africa: Redefining Roles and Responsibilities of Public and Private Actors in the Mining Sector." *Resources Policy*, Special issue: 'Corporate Social Responsibility in the Extractive Industries: Experiences from Developing Countries' 37, no. 2 (2011): 138–143.

Campbell, B. "Revisiting the Reform Process of African Mining Regimes." *Canadian Journal of Development Studies* 30, no. 1–2 (2010): 197–217.

Chavagneux, C. "Peut-on maîtriser la mondialisation? Une introduction aux approches d'économie politique internationale." [Can We Control Globalization? An Introduction to Approaches in International Political Economy] *Économies et Sociétés* 34, no. 4 (1998): 25–68.

Clausen, F. M. L. Barretto, and A. Attaran. "Property Rights Theory and the Reform of Artisanal and Small-Scale Mining in Developing Countries." *Journal of Politics and Law* 1, no. 4 (2011): 15–26.

Clifford, M. J. "Potential Repercussions of a Mercury Ban on the Artisanal and Small-scale Gold-mining Sector: A Viewpoint." *International Journal of Environment and Pollution* 41, no. 3–4 (2010): 229–241.

Communities and Small-scale Mining (CASM). "Information." n.d., http://www.casm.org.

D'Angelo, L. "Who Owns the Diamonds? The Occult Eco-nomy of Diamond Mining in Sierra Leone." *Africa* 84, no. 2 (2014): 269–293.

De Failly. D., Bulakali Ntakobajira, Z., and L. Bahimba Shonja. "Traceability of Local Mining Flows: Governance Issues and Poverty Alleviation Challenges in the Artisanal Sector of the Democratic Republic of the Congo (DRC)." In *Modes of Governance and Revenue Flows in African Mining*, International Political Economy Series, edited by B. Campbell, 164–222. Basingstoke: Palgrave Macmillan, 2013.

De Theije, M., and M. Heemskerk. "Moving Frontiers in the Amazon: Brazilian Small-scale Gold Miners in Suriname." *European Review of Latin American and Caribbean Studies* 87 (2009): 5–25.

Doucouré, B. *Des pierres dans les mortiers et non du maïs! Mutations dans les villages aurifères du sud-est du Sénégal* [Stones in the Mortars Rather than Corn]. Dakar: CODESRIA, 2015.

Gagné-Ouellet, S. "Regulatory Framework Review and Mining Regime Reform in Mali: Degrees of Rupture and Continuity." In *Modes of Governance and Revenue Flows in African Mining*, International Political Economy Series, edited by B. Campbell, 47–100. Basingstoke: Palgrave Macmillan, 2013.

Ferring, D., H. Hausermann, and E. Effah. "Site Specific: Heterogeneity of Small-scale Gold Mining in Ghana." *The Extractive Industries and Society* 3, no. 1 (2016): 171–184.

Fisher, E. "Occupying the Margins: Labour Integration and Social Exclusion in Artisanal Mining in Tanzania." *Development and Change* 38, no. 4 (2007): 735–760.

Hentschel, T., F. Hruschka, and M. Priester. "Global Report on Artisanal and Small-scale Mining." *Report Commissioned by the Mining, Minerals and Sustainable Development of the International Institute for Environment and Development* 20, no. 8 (2002). http://www.iied.org/mmsd/mmsd_pdfs/asm_global_report_draft_jan02.pdf.

Hilson, G. "Artisanal Mining, Smallholder Farming and Livelihood Diversification in Rural Sub-Saharan africa: an introduction." *Journal of International Development* 23, no. 8 (2011): 1031–1041.

Hilson, G., and S. Pardie. "Mercury: An Agent of Poverty in Ghana's Small-scale Gold-mining Sector?" *Resources Policy* 31, no. 2 (2006): 106–116.

Hilson, G., and A. Ackah-Baidoo. "Can Microcredit Services Alleviate Hardship in African Small-scale Mining Communities?" *World Development* 39, no. 7 (2011): 1191–1203.

Hilson, G. M., and S. Van Bockstael. "Diamond Mining, Rice Farming and a 'Maggi cube': A Viable Survival Strategy in Rural Liberia?" *Journal of International Development* 23, no. 8 (2011): 1042–1053.

Hilson, G., and G. Okoh. "Artisanal Mining in Ghana: Institutional Arrangements, Resource Flows and Poverty Alleviation." In *Modes of Governance and Revenue Flows in African Mining*, International Political Economy Series, edited by B. Campbell, 138–163. Basingstoke: Palgrave Macmillan, 2013.

Hilson, G., and J. McQuilken. "Four Decades of Support for Artisanal and Small-scale Mining in Sub-Saharan Africa: A Critical Review." *The Extractive Industries and Society* 1, no. 1 (2014): 104–118.

Hinton, J. J. "Communities and Small-scale Mining: An Integrated Review for Development Planning." *Communities and Small-Scale Mining (CASM) Initiative*. World Bank, 2015: 195.

Hinton, J. J., M. M. Veiga, and C. Beinhoff. "Women and Artisanal Mining: Gender Roles and the Road Ahead" Chap. 11 in *The Socio-economic Impacts of Artisanal and Small-scale Mining in Developing Countries*, edited by G. Hilson. Netherlands: A.A. Balkema, Swets Publishers, 2003: 149–188. https://circle.ubc.ca/bitstream/handle/2429/35920/ubc_2011_fall_hinton_jennifer.pdf?sequence=5.

ICMM (International Council on Mining and Metals). "Artisanal and Small-scale Mining." ICMM, 2015. http://www.icmm.com/page/84136/artisanal-and-small-scale-mining.

ILO (International Labour Organisation). *Social and Labour Issues in Small-scale Mines. Report for Discussion at the Tripartite Meeting on Social and Labour Issues in Small-scale Mines*. Geneva: International Labour Organization, Sectoral Activities Program, International Labour Office, 1999.

Jennings, N. S. "Addressing Labour and Social Issues in Small-scale Mining." In *The Socio-economic Impacts of Artisanal and Small-scale Mining in Developing Countries*, edited by G. Hilson, 151–160. Rotterdam: A.A. Balkema, 2003.

Jønsson, J. B., P. W. U. Appel, and R. T. Chibunda. "A Matter of Approach: The Retort's Potential to Reduce Mercury Consumption within Small-scale Gold Mining Settlements in Tanzania." *Journal of Cleaner Production* 17, no. 1. (2009): 77–86.

Levin, E. A., and L. Gberie. Dealing for Development? A Study of Diamond Marketing and Pricing in Sierra Leone, Diamond Development Initiative. DDI, 2006. http://www.estellelevin.com/wp-content/uploads/2013/12/DDI_A_Study_of_Diamond_Marketing_and_Pricing.pdf.

Luning, S. "Liberalisation of the Gold Mining Sector in Burkina Faso." *Review of African Political Economy* 35, no. 117 (2008): 387–401.

Maconachie, R. "RE-agrarianizing Livelihoods in Post-conflict Sierra Leone? Mineral Wealth and Rural Change in Artisanal and Small-scale Mining Communities." *Journal of International Development* 23, no. 8 (2011):1054–1067. http://tinyurl.com/guct7jz.

Maconachie, R., and T. Binns. "'Farming Miners' or 'Mining Farmers'?: Diamond Mining and Rural Development in Post-conflict Sierra Leone." *Journal of Rural Studies* 23, no. 3 (2007): 367–380.

McQuilken, J., and G. Hilson. 'Artisanal and Small-scale Mining in Ghana: Background Research to Inform an Action Dialogue.' *International Institute for Environment and Development* (2016): 1–43.

Naito, K., F. Remy, and J. P. Williams. *Review of Legal and Fiscal Frameworks for Exploration and Mining*. London: Mining Journal Books, 2001.

Ndimubanzi Ngoroba, E. "Décalage entre les textes légaux régissant le secteur minier et la réalité sur le terrain [The Gap between the Legal Texts Governing the Mining Sector and the Reality on the Ground]." In *Les Minerais de 'Sang': Un secteur économique criminalisé à l'Est de la R.D.Congo* ['Blood' Minerals: A Criminalised Economic Sector in the East of the Democratic Republic of the Congo], Dossier 08bis, 16–20. Goma: Pole Institute, (2010).

Perks R. "The Inclusion of Artisanal and Small-Scale Mining in National Legislation: Case Studies from Sub-Saharan Africa." Paper presentation at the 59th Annual meeting of the Rocky Mountain Mineral Law Institute, Spokane, WA, July 18–20, 2013.

Siegel, S., and M. M. Veiga. "Artisanal and Small-scale Mining as an Extralegal Economy: De Soto and the Redefinition of "formalization"." *Resources Policy* 34 (2009): 51–56.

Spiegel, S. "Governance Institutions, Resource Rights Regimes, and the Informal Mining Sector: Regulatory Complexities in Indonesia." *World Development* 40, no. 1 (2012): 189–205.

Strange, S. *States and Markets*. 2nd ed. London: Continuum, 1994.

Strange, S. *The Retreat of the State: The Diffusion of Power in the World Economy*. Cambridge: Cambridge University Press, 1996.

Strange, S. "International Political Economy: Beyond Economics and International Relations." *Économies et Sociétés* 34, no. 4 (1998): 3–24.

Szablowski, D. *Transnational Law and Local Struggles: Mining, Communities and the World Bank*. Oxford: Hart Publishing, 2007.

Teschner, B. A. "Small-scale Mining in Ghana: The Government and the Galamsey." *Resources Policy* 37 (2012): 308–314.

UNECA (United Nations Economic Commission for Africa). "Compendium on Best Practices in Small-scale Mining in Africa." Working paper, December 2002. Addis Ababa: Economic Commission for Africa, http://repository.uneca.org/handle/10855/5447.

UNECA. "Viable SME Mining." Presentation to workshop on small and medium sized enterprises and the development of the extractive industry, Johannesburg, December 2009.

UNECA and AU. *Minerals and Africa's Development. The International Study Group Report on Africa's Mineral Regimes*, Addis Ababa: Economic Commission for Africa, 2011.

Van Bockstael, S. "The Persistence of Informality: Perspectives on the Future of Artisanal Mining in Liberia." *Futures* 62 (2014): 10–20.

Verbrugge, B. "The Economic Logic of Persistent Informality: Artisanal and Small-scale Mining in the Southern Philippines." *Development and Change* 46, no. 5 (2015): 1023–1046.

Verbrugge, B. "Undermining the State? Informal Mining and Trajectories of State Formation in Eastern Mindanao, Philippines." *Critical Asian Studies* 47, no. 2 (2015): 177–199.

Verbrugge, B., and B. Besmanos. *Undermining the Myths about Small-scale Mining*. Manila: Bantay Kita, 2015.

World Bank. *Strategy for African Mining*, World Bank Technical Paper No. 181, Africa Technical Department Series, Mining Unit, Industry and Energy Division, Washington, DC: World Bank, 1992.

Yakovleva, N. "Perspectives on Female Participation in Artisanal and Small-scale mining: A Case Study of Birim North District of Ghana." *Resources Policy* 32 (2007): 29–41.

The politics of artisanal and small-scale mining in Mongolia

Pascale Hatcher

ABSTRACT
Large-scale mining has become the cornerstone of Mongolia's development and poverty reduction narrative. Building on political economy insights, this paper explores the politics of artisanal and small-scale mining (ASM) in the competing context of the fast-pace development of large-scale mining in Mongolia. Questions are raised about the actors involved in promoting an 'investment-led model' which has privileged large-scale activities. Such context sheds light on the marginalisation of ASM activities in Mongolia, a sector which, despite being illegal until 2010, has continued to support 20% of Mongolia's rural workforce.

> You can't just stop artisanal mining. ASM will continue as long as poverty persists. (Mongolian government official, SAM conference[1])

Introduction

Mongolia's mining riches are astounding. It is estimated that the country rests on 6000 deposits of 80 different minerals, notably coal, copper, gold, rare earth and iron ore.[2] Spearheaded by international donors, Mongolia's mining regime has been highly successful in attracting foreign large-scale investments. In fact, the country is currently the construction site of what are expected to become two of the biggest mines on Earth.

While in recent years, large-scale mining has become the cornerstone of Mongolia's development and poverty reduction narrative, there are growing tensions emerging from the severe socio-environmental impacts linked to the rapid expansion of the sector. Crucially, if all eyes are currently on the large-scale sector, artisanal and small-scale mining (ASM), despite being illegal until 2010, continues to support 20% of Mongolia's rural workforce and to account for nearly 10% of revenues in the country. With the competing context of the fast-paced development of large-scale mining in the background, the paper builds on political economy insights to explore the politics of mining as a development strategy in Mongolia and asks why ASM has continued to be marginalised despite its important economic contribution.[3]

This paper is divided into four sections. In the first part, the paper presents the making of Mongolia's current mining regime, one that was highly successful in its objective of developing the industrial side of the sector under a particular 'investment-led' framework. The second section of this paper turns to ASM and analyses how the sector has developed alongside the country's large-scale sector. In turn, the paper discusses the process by which ASM has systematically been delegitimised and marginalised. In the last section of the paper, the current debate over the need to rethink the country's mining regime is analysed.

The making of Mongolia's large-scale mining sector

While industrial mining activities took place across Mongolia during its 70 year long socialist era, Mongolia has only recently begun to position itself as one of the world's key mineral producers. The foundations of the legal framework, which has propelled the country to its current status, were drawn up in the early 1990s, amidst the country's rapid transition towards a market economy. This section of the paper raises questions relating to the actors involved in promoting specific *norms* for and *legitimising* a neoliberal mining governance in the country. Such context sheds new light on the marginalisation of ASM activities in the country.

Sponsored by a handful of multilateral organisations, the initial 'shock therapy' saw the liberalisation of prices, the removal of trade restrictions and the privatisation of state enterprises.[4] Amidst this liberalisation, large-scale mining was quickly identified as a key sector to generate the country's much needed revenues – the country's budget was no longer supported by the Soviet Union. In the mid-1990s, alongside most of the other resource-rich countries in the region, Mongolia embarked on a reform process that led to the adoption of a new mining legislation.[5] The 1997 Mineral Law, and later the Foreign Investment Law (2002) were closely modelled on specific neoliberal norms and procedures, as advocated by the World Bank and the Asian Development Bank.[6] These new laws sought to encourage foreign investments in the country's large-scale sector. The World Bank[7] observed that:

> The Bank has been supporting and nurturing the development of the regulatory framework for private sector led mining sector exploration and development for more than 10 years. During this time the focus of activities has been on establishing the legislative and fiscal regimes to apply to mining […]. This work culminated in the passing of the 1997 Mineral Law and adoption of a reasonably competitive fiscal regime for the sector.[8]

Beyond the specific case of Mongolia, and as also argued by Campbell in this collection, the importance of the World Bank's overarching influence on the liberalisation and deregulation of mining sectors across poor indebted countries over the better part of the last three decades should not be understated.[9] It is worth noting that the Extractive Industries Review (EIR), which was established in 2001 to independently evaluate the World Bank Group's (WBG) involvement in extractive industries, estimated that under the distinct leadership of the Bank no less than 100 countries had reformed their laws, policies and institutions during the 1990s.[10] The EIR further stressed that 'in line with WBG advice,' this new legislation, which was designed to ensure the protection of capital and to promote investment, led many developing countries to experience an investment boom in their mining, oil and gas sectors.[11]

Along the lines of the World Bank's recommendations, resource-rich countries adopted new legal frameworks prioritising a specific set of neoliberal norms and procedures geared to develop mining within a specific governance framework. Gagné-Ouellet summarises these norms[12]:

- priority given to the private sector for mining development;
- priority given to mining over other types of territorial use;
- priority given to an exportable resource over other mineral resources;
- priority given to the industrial sector over artisanal and small mines; and
- guarantees protecting mining rights.

In line with such neoliberal norms, Mongolia's new mining regime sought to encourage foreign investors to participate in the exploration and *large-scale* development of the country's mineral riches.[13] Unsurprisingly, therefore, the 1997 Mining Law was regarded at the time as one of Asia's most attractive mining laws to foreign investors.[14] The new Law prioritised the transparency, security and transferability of property rights. Additionally, its low taxation and royalty burdens were particularly enticing to foreign investors.[15]

Crucially, and in accordance to multilateral financial organisations' guidelines for the sector, the role of the Mongolian state shifted from an 'ultimate land owner' to a manager/regulator of the sector.[16] These new roles contributed to a profound shift in the power relations between the actors involved in the sector. As discussed by Campbell in this collection, this 'stringent retrenchment of state' led to a new delineation between public and private spheres of authority, transferring what were previously considered public responsibilities to private actors.[17] Albeit Campbell here is discussing the case of Africa; her observations resonate with the case of Mongolia. As argued later in this contribution, this retreat of the state accounts for several of the failures by the Mongolian government to address some of the most severe socio-environmental impacts attached to the large-scale mining sector.

The neoliberal shift further accounts for a significant transfer of land rights to mining corporations, away from local nomadic people. Historically, and still today, agriculture has accounted for only a minimal portion of land use in Mongolia – approximately 1% – leaving livestock grazing to account for the overwhelming share of land use.[18] As plainly observed by Lahiri-Dutt and Dondov: 'Clearly, private ownership of rangelands has never existed in Mongolian history. Rangelands have always been state-owned property used in common by herders or herder groups.'[19] However, despite the importance of nomadism for the livelihood of herders, who represent approximately 30% of the population,[20] the 1990s saw the adoption of neoliberal laws expressly privileging large-scale mining over any other type of territorial use, thus marginalising both herders and, as discussed in the following section, the emerging mass of people turning towards ASM.

The liberalisation brought forth by the 1997 Mineral Law, as well as the Gold Programme which was being actively promoted by the government, marked the beginning of what was to quickly become a significant industrial 'mining boom' across the country. Urantsooj, president of the Mongolian Non-Governmental Organisation (NGO) Center for Human Rights and Development, observes that the fast pace of issuance of mining licences at the time, notably for the gold sector, has meant that 'almost the whole country was covered.'[21] Although revisions to the 1997 Mining Law in 2006 came to challenge some of its most liberal provisions, the country quickly became one of the world's key mineral producers. By 2012, Mongolia had reached the seventh position in terms of world mineral and ore production,[22] making it the centre of what the *Financial Times* referred to as an 'investment bonanza.'[23]

It is illustrative to note that the Gobi desert has been the construction site of what is projected to become the world's largest coal mine (Tavan Tolgoi), and the third-largest copper and gold mine (Oyu Tolgoi, or OT). The Government of Mongolia holds 34% of common stocks in OT while the remaining 66% of shares are in the hands of the foreign-owned company Turquoise Hill Resources. The influential British Australian mining corporation Rio Tinto owns a majority stake in Turquoise Hill Resources, which makes it the largest foreign investor in the country. In December 2015, the World Bank's International Financial Corporation and the Multilateral Investment Guarantee Agency committed US$2,2 billion in loan and guaranteed for the second phase of OT mine, which is now the largest mine in the country.[24]

Exceeding all expectations, Mongolia's economy was lifted by the mining wave and by 2011, it had jumped to a staggering 17.5%, making it the fastest growing economy in the world at the time – nearly twice as fast as China's.[25] From contributing close to 13% of the country's gross domestic product (GDP) in 2002, mining accounted for 30% of the country's GDP in 2012.[26] Spearheaded by its booming large-scale sector, the size of Mongolia's economy tripled between 2006 and 2013.[27]

Amidst demand from China and the commodity boom, the neoliberal norms drove the expansion of the sector and came to legitimise an 'investment-led' strategy for the fast-paced development of the mining sector. Here, the large-scale sector found its legitimacy not only via the approbation of international donors and other organisations,[28] but also with the socio-economic promises attached to the booming sector. These were closely linked to tax contribution and employment creation. However, and as discussed in the following section of this paper, this investment-led model further served to delegitimise ASM in the country, despite the sizeable contribution of the ASM sector to poverty reduction.

The rise of the Ninjas

While the 1997 Mining Law decidedly entwined the fate of the country's economy to the industrial sector, it is telling to note that the Law completely ignored the very existence of ASM – the Law does not mention it once. However, and as indicative of the lack of legal status of ASM activities in many countries across the Global South, despite this de facto illegal status, the sector had already begun to flourish in Mongolia.[29] However, it should be noted that in sharp contrast to many resource-rich countries, ASM in Mongolia is a very recent phenomenon.[30] It emerged as a direct aftermath of the 'shock therapy' packages embraced by the country in the early 1990s, amidst the country's transition towards a market economy.[31] The socio-economic impacts of the market reforms were sizeable – the GDP fell by approximately 20% at that time.[32] While unemployment had not been a problem before 1990, it had reached 8.7% by 1994, and amidst the considerable waning of the country's social welfare system, poverty levels rose steeply.[33] The rise in poverty contributed to a mass migration towards the capital, around which coal began to be mined informally.[34]

In addition to the country's bumpy turn towards the market economy, a series of extreme weather conditions contributed to the sudden appearance of ASM.[35] Between 1999 and 2002, Mongolia saw three consecutive extremely snowy and cold winters known as *zuds*, which seriously impacted livestock.[36] Between 7 and 11 million head of cattle and other livestock perished in the aftermath of such *zuds*,[37] pushing rural communities into 'extreme poverty and economic desperation.'[38] The majority of these herders further moved to the capital and/or turned to ASM as an economic activity of last resort.[39]

It was however with the rise of gold prices that ASM established itself as a *permanent* activity in Mongolia.[40] Since the mid-2000s, the dwindling legal gold supplies and the spike in black market demand from neighbouring China have further contributed to heightened profits for ASM of gold in the country.[41] As observed by the Agency for Development Cooperation,[42] with the ensuing gold rush, the unemployed urban population joined the ex-herders in becoming 'ninjas' – a term colloquially used to refer to small-scale and artisanal miners operating in the country.[43]

The overwhelming majority of ASM miners – between 80 and 90% – mine gold (both primary and placer),[44] while the remainder mine for fluorspar and coal,[45] as well as a small amount of semi-precious stones, tungsten and petrified wood. Due to Mongolia's harsh winters, ASM is generally a seasonal occupation – although by no means exclusively.[46] Men and women – the latter represent approximately a third of the ASM workforce[47] – involved in such activities include herders, farmers, students, factory workers, and even white-collar workers such as government officials.[48] While on average, the sector pays much better than the Mongolian minimum wage, income from ASM remains particularly intermittent and volatile.[49]

The sudden development of the phenomenon, as well its scale, left the country's authorities in a state of shock and denial.[50] Since its initial outbreaks in 1995–1996, ASM has spread to all but three of the country's 21 provinces. While until recently, authorities have resolutely tried to ignore the sector's very existence, as discussed in the following section, it is compelling to note that it has continued to expand alongside – and benefit from – the prioritisation of the industrial sector.

ASM in Mongolia remains controversial, a fact reflected in the attempts to estimate the number of the country's miners.[51] While some figures point towards 35,000 operating in the country, the actual figure is most probably closer to 100,000 people.[52] Taking into account the fact that Mongolia's total population is three million, the sector's significance should not be understated. In fact, in rural areas, ASM is the largest generator of income after agriculture and accounts for 20% of the total workforce.[53] Moreover, these numbers do not account for the fact that each 'ninja' is economically supporting three to four people,[54] nor for the employment created by investment of mining income (such as petty commerce).[55] Today, up to 400,000 people indirectly depend on the sector (13% of the country).[56]

ASM as a nuisance

Despite its substantial contribution to the local economy, ASM has until recently been perceived as a nuisance. High observes that for Mongolians living in Ulaanbaatar, the capital, ninja mining is commonly regarded as 'a blemish, if not an embarrassment.'[57] The widespread disapproval of ASM activities in the country has overwhelmingly sprung from the sector's environmental legacy, notably its impact on water and land, not withstanding the widespread use of mercury.[58] It is however important to stress that to a large extent, ASM's severe environmental footprint derives from the fact that the sector was illegal until recently. Purevjav observes that miners have been struggling to implement environmental friendly practices in relation to rehabilitation of degraded land due to their lack of access to knowledge, financial resources and technological capacity.[59]

Beyond such public perception, and as further argued by Campbell in this collection, the appearance of a clear dichotomy between ASM and the large-scale sector is somewhat removed from empirical reality.[60] This point comes to light when analysing the tensions between ninjas and their two main opponents: herder communities and the large-scale sector.

Tensions between herders and miners have grown to be increasingly serious, with complaints linked to the deterioration of subsoil; livestock falling into the holes dug by miners; dried up rivers and springs; and water pollution. However, the herders/miners dichotomy oversimplifies the reality on the ground as ninjas are often ex-herders themselves or even continue to herd livestock while complementing their revenues with mining activities. Lahiri-Dutt and Dondov further observe that the analysis should take into account the fact that herders who transitioned to ASM tended to have small herds, while those with larger herds were able to 'withstand the onslaught of change.'[61] The authors thus make the important point that the conflicts between herders and miners 'can be interpreted largely as class conflicts rather than the vulnerability of a livelihood from the activities of another.'[62]

Increasing tensions have also been noted between ASM miners and the industrial sector. In a 2006 working paper, the World Bank observed that while initially large-scale operators tolerated the presence of informal miners if their activities were confined to the waste tailings, tacit understandings quickly broke down.[63] The Swiss Agency for Development Cooperation (SDC) notes that: 'The industrial mining sector has long seen ASGM [Artisanal small-scale gold mining] as an annoyance and depicted artisanal miners as villains, irresponsibly "destructing" [*sic*] mineral deposits and the environment, and opposing against modern mechanized mining.'[64] However, this apparent dichotomy again hides another reality on the ground:

> In areas where LSM [large-scale mining] works, ASGM [Artisanal and Small-scale Gold Mining] is often either a traditional activity or artisanal miners had discovered the gold deposits. There are abundant cases – where these deposits turned out to be outcrops of a larger mineralization – where ASGM was subsequently displaced by LSM companies. Concession areas for LSM are often significantly larger than the area actually required for the mining operation, sometimes covering thousands of square kilometres. Such areas limit the possibilities of artisanal miners to do ASGM formally, and even on community land where ASGM had been a traditional activity, miners are often considered illegal and criminal invaders.[65]

As further discussed by Campbell in this collection, while large-scale mining activities can deliver economic growth, the 'investment-led model' has often failed to translate into economic benefits at local levels.[66] In contrast with industrial mining, in ASM, 'money first enters at the household level and begins to circulate from the bottom up, from the local to the regional and national economies.'[67] As observed by Mrs. Saruul, the head of a local ASM formal partnership (association) in the Tsagaan tsakhirt site (gold mining):

> I believe that small-scale mining has more potential to address poverty reduction compared to large-scale mining. I strongly believe so for two reasons. Firstly, ASM affects the lives of many in a direct way, meaning it can reduce poverty directly. Secondly, large-scale mining uses machinery which causes much more environmental destruction than ASM.[68]

Further, while promises of employment are key to the narrative promoting an investment-led model, it is interesting to note that according to the National Statistical Office of Mongolia, only 8268 people were engaged in mining of metal ores in 2012, of a total of 19,217 people employed in the mining and quarrying sector and 61,661 people in the entire industrial division.[69] This is a far cry from the 100,000 ASM workers, and the 300,000 people being indirectly supported economically by ASM activities. This is not to downplay the quality and security of the much-coveted jobs in the large-scale sector but rather to emphasise the false premises attached to the overwhelming focus on an investment-led mining regime in terms of employment.

Further to this, and despite public opinion being focussed on the environmental damage in the ASM sector, the investment-led model has registered a serious environmental footprint. Mongolia's Mineral Resources Authority admits that since the beginning of the mining boom, 4000 ha of land have been adversely affected by mining – 500 mining sites affecting 56 towns in 15 different provinces[70] and observes that:

> Companies neglected environmental standards, and fared poorly on rehabilitation. Use of poor technology and obsolete equipment, the tendency of placer gold mines to be located in river basins, the use of river water for gold extraction and putting waste into rivers resulted in numerous reported cases of pollution and some rivers drying up. This, combined with the low regulatory enforcement capacity of the government, led to environmental problems which started affecting rural livestock herders as pastures and water resources became scarce. Many civil society groups, especially at the grassroots level, emerged in reaction to environmental concerns around mining, and ultimately some of these groups went beyond environmental issues in their quest to bring about more responsible mining practices.

Fed by such tensions and a 'widespread public opposition,'[71] authorities have treated the sector as a 'temporary annoyance.'[72] As such, most governmental initiatives have remained framed within the over-optimistic idea that if kept illegal, the sector would simply disappear.[73] It is illustrative to note that throughout the 2000s, and to a certain extent till today, the terms 'ninjas,' 'informal' and 'artisanal' all appear to mean 'illegal' mining.

In such an illegal environment, ASM miners have been living on the edge of society, lacking security, knowledge and skills to sustain their activity, as well as lacking much needed access to health care and social protection. This closely resonates with Hall, Hirsch and Murray Li's work on the concept of exclusion,[74] which they see as structured by power relations, whether related to regulation, force, the market or legitimation.[75]

In the Mongolian case, and as argued in this contribution, one can clearly observe that regulations have indisputably favoured the industrial sector. In a de facto state of illegality, ASM workers have been the subjects of use of force by both private and public security forces. The SDC report instances of 'police and military raids, direct confrontation, arrests and trials with – in some cases – rampant violations of human rights.'[76] In terms of structured power relations linked to the market, ninjas have a limited access to information and means to sell gold via legal channels. Here, the 'ger shop,'[77] or the mill or truck owners have become the only accessible source of financial service for miners, as well as a convenient venue to sell the gold – mining sites are often far from financial institutions and Ulaanbaatar. Most middlemen then illegally sell the gold to Chinese or South Korean markets.[78] In 2002, in order to curb these illegal channels, the Mongol Bank (the central bank) commenced purchasing gold at the world price from ASM directly and from gold buyers as private individuals. However, illegal channels have remained rampant.

Finally, and keeping with Hall, Hirsch and Murray Li's approach to the concept of exclusion,[79] while large-scale mining activities in Mongolia have rapidly become 'legitimate,' ASM has been systematically marginalised. In addition to its inability to mirror the scale of the economic promises of the industrial sector, ASM has lacked legitimacy in public opinion due to its historical environmental footprint across the country. Unsurprisingly therefore, the government of Mongolia has long been cautious in proposing any legislation in regard to the legalisation of the sector 'for fear of being perceived as supporting a sector that remains so unpopular with the public.'[80]

However, in light of the importance played by ASM in rural economic development, employment creation and poverty reduction, the Mongolian state, in collaboration with NGOs and international organisations, began to study regulatory options in the early 2000s.[81] This is discussed in the following section.

The legalisation of ASM

The momentum which brought the authorities to begin discussing ASM inscribed itself in important changes across the large-scale mining sector in the mid-2000s. In 2006, an amendment to the 1997 Mining Law curbed some of its most liberal provisions. The amendment sought to increase tax rates and state ownership over minerals, stipulating that in the cases of exploitation of minerals of 'strategic importance,' the State would claim up to 34% (Art. 5.5).[82] While the World Bank was quick to condemn the adoption of the amendments as a 'pushback' by the state on private sector led development,[83] the changes conveyed the sense of unease regarding what was widely perceived as the over-liberalisation of the large-scale sector and its correlated socio-economic and ecological impacts.

Amidst these changes in the overall mining sector, the role of ASM gradually began to be revisited. Initial regulations were tentatively introduced as early as 2001[84] but the discussions over the official legislation faced insurmountable implementation problems over the better part of the following decade – notably due to a lack of adequate consultation of the miners themselves and therefore, of the lack of understanding of the sector by the authorities.[85] In December 2010, the formal approval of Resolution 308 marked the official legalisation of ASM.[86]

The move was hailed as 'the first progressive step toward formalisation of SSM.'[87] With the 2010 Law, the government was recognising the rights of artisanal miners to work, mine land and organise themselves into official 'partnerships.'[88] The legislation stipulates that miners must organise in 'unregistered partnerships' of no less than five members in order to engage in mining, which are regulated by local administrations and limited to areas not yet licenced for exploration and exploitation.[89] In other words, the new legal provision gives artisanal miners the opportunity to scale up their activities into more formalised small-scale mining, to have a plot of land upon which to legally mine – provided that: (i) deposits are of no commercial value for industrial exploitation; (ii) deposits are already exploited waste deposits that have been abandoned and left as waste; and (iii) the company and local authorities have mutually agreed to allow ASM work to take place.

A year after the enactment of the legalisation, 317 new partnerships had been established.[90] By December 2015, 7325 people had formally registered in the sector, 30% of which were women.[91]

An important factor contributing to the momentum for the recent changes in ASM was the fact that despite operating in complete illegality, ASM was already supporting 15% of the country's entire population – albeit indirectly. As such, the local and national government authorities were brought to realise that if harnessed, the sector was promising to sizably contribute to tax revenues.

The development of the policy built on several consultations held in 2009–2010 with more than 400 stakeholders, including Members of Parliaments, central and local government officials, miners' representatives, NGOs and the artisanal miners themselves.[92] Crucially, the Swiss Agency for Development Cooperation (SDC) was a pivotal actor in fostering a change in attitude towards the sector.[93] The SDC's 'Agency's Sustainable Artisanal Mining Project' (SAM), which began in 2005, played a fundamental role in introducing the concept of sustainable artisanal mining into public and policy dialogue[94]:

> In Mongolia, in absence of a gremial ASGM organization, and a historically stronger top-down governance approach, SAM and particularly its national counterpart had to take the lead. Based on broad and participatory stakeholder consultation, several drafts of an ASM law were presented between 2005 and 2007 and discussed in Parliament; [...]. The SAM project further supported the law-making process until an ASM Law was finally passed in 2010.[95]

The creation of local NGO 'Mongolian Artisanal Miners' United Umbrella Association' (MASM) in May 2013 further marked a new step in the formalisation of ASM activities in the country. Established by Mongolian miners themselves, the NGO represents over 7000 ninjas. Its goal is 'to develop socially and environmentally responsible artisanal and small-scale mining that is recognised by the society by protecting member organisations rights and improving livelihoods of miners.'[96]

While important challenges remain, the changes have led to substantial gains for some of the miners now organised in partnerships. With the newly found land rights, the legalisation of the sector gives miners the confidence to invest in technological improvements and workplace safety, resulting in 'better health, lesser accidents, higher productivity and secured livelihoods.'[97] It has also encouraged much needed training for the sector, as 90% of the workers are 'non-professionals.'[98] Furthermore, access to social services has become a noticeable benefit of the legalisation of the sector. The Law stipulates that the *Soum* [district] governor is responsible for taking 'measures to provide health care and social welfare services to small-scale miners, and access to schools and kindergartens to their children.' For instance, in 2015 alone, there was a 119% increase in health and social insurance enrolment by artisanal miners, 40% of which were women.[99]

In terms of environmental practices, the formalisation of the sector and the availability of mercury-free technological alternatives have led to what the SDC refers to as 'excellent progress.'[100] As observed by Mrs Saruul, the head of a newly formed local ASM partnership organised as an NGO in the Bayankhongor *aimag*[101] (Tsagaan Tsakhirt) site, partnerships transformed ninjas' own attitudes over the years:

> People [miners] were only concerned about how to make ends meet. After finishing digging in one place, we simply moved to new land without taking environmental remediation into account. In fact, we simply did not know about the side effects of illegal mining on both society and environment.[102]

Already between 2013 and 2014, ASM gold sales rose by 115.2%, validating what the SDC refers to as 'the immense economic potential of ASM when systems and mechanisms for formalization are in place to fully realize that potential.'[103] Furthermore, amidst the current economic downturn caused by the decrease in world prices and the decline of China's demand for minerals, the legalisation process of ASM has allowed for the sector to make a tangible contribution to the economy.[104] In 2015, the sector sold 6.7 tonnes of gold to the Bank of Mongolia – almost half (46.3%) of total gold-mining sales in the country.[105] This represents USD 245 million in foreign exchange earnings and USD 6.1 million in royalties.[106] Unsurprisingly therefore, the public and political perceptions of ASM have improved considerably in the last year.[107]

Despite such successes, however, important challenges remain, both in terms of the formalisation process, and in making sure that the gains of the formalisation continue to be advantageous for the ASM community. As of 2015, approximately 89% of ASM activities remain informal.[108] According to the SDC, this is mostly due to miners' reluctance to pay taxes and a lack of access to viable land for the miners.[109] However, in addition to a range of projects currently unfolding to further support the formalisation of the sector,[110] the Mineral Resource Authority of Mongolia's recent increase in allocation of ASM mining land is expected to substantially expand the possibilities for formalisation in the sector.[111]

Conclusion

Promoted by international donors, the investment-led mining framework adopted by Mongolia was highly effective in attracting large-scale investments to the country. However, and as argued in this paper, while ASM has long been overshadowed by the socio-economic promises of this investment-led regime, there is a need to revisit the origins of the country's regulatory mining framework to understand the legitimisation process which led to the adoption of such framework.

This process sheds light on the existence of relations of power – relations in which external actors continue to play a central role – which has constrained changes in the neoliberal norms firmly embedded in the regime. In other words, these relations of power inform the conditions which would permit ASM to become an integral part of the discussion over the role of mining activities in the country's economy – and society a whole – rather than a marginalised sector.

However, in Mongolia and elsewhere, governments have recently begun to challenge some of the more liberal norms which have historically been promoted within an investment-led model. Amidst such changes, a handful of NGOs, bilateral agencies as well as small-scale miners have increasingly been pressuring the government authorities to acknowledge the socio-economic potential of such activities. While the ensuing changes have remained limited, the exponential economic contribution of small-scale gold mining to the national budget is impressive, and so is the increasing attention paid to better environmental practices in the sector. Finally, and as also noted by Campbell in her contribution to this collection, there is a clear need to link large- and small-scale mining in a manner which articulates rather than opposes activities within the sector as a whole and crucially, in a manner that prioritises the entire sector's contribution to the welfare of Mongolian people.[112]

Disclosure statement

No potential conflict of interest was reported by the author.

Funding

This research was supported by the Program to Support General Research Activities (*Kiban-kenkyu*), Ritsumeikan University and the field research in Mongolia (2015) was supported by a Grant-in-Aid for Scientific Research (*Kakenhi*) from the Japanese Government; Japan Society for the Promotion of Science [grant number 15H05388 (2015-17)].

Acknowledgement

I am grateful to my research assistant, Suzie Boulanger (Centre interdisciplinaire de recherche en développement international et société, Université du Québec à Montréal) for her valuable contribution. I would also like to thank Bonnie Campbell and Chris Huggins for their valuable comments on the paper.

Notes

1. Villegas, "Ninja Miners and Rural."
2. In addition to these minerals, Mongolia is believed to have reserves of asbestos, bismuth, clays, copper, diamond, gemstones, graphite, gypsum, lead, limestone, magnesium, molybdenum, nickel, petroleum, phosphate, platinum group metals, rare earths, salt, sand and gravel, silica, talc, tin, tungsten, uranium, zeolite and zinc. Wu, "Mineral Industry of Mongolia," 15.1.
3. The argument in this paper mostly builds on secondary data analysis. It should be noted that due to its de facto illegal nature since it appeared in the early 1990s, data on the topic are sparse – and remains so despite its recent legalisation. The Swiss Agency for Development and Cooperation must be acknowledged as an important source of data on the topic. Additionally, some of the material is corroborated by field visits held between 2012 and 2015 in Ulaanbaatar where government officials, NGOs and international organisation officers, and mining company employees were interviewed.
4. These austerity packages were first adopted in 1991 and later, in 1994.
5. For instance, Myanmar adopted a new mining law in 1994, Philippines in 1995, Vietnam and Kazakstan in 1996, China, Laos and Kyrgyzstan in 1997. On the topic, see Naito et al., "Mineral Projects."
6. The 1997 Minerals Law regulated relations with respect to exploration and mining of all types of mineral resources except water, petroleum and natural gas from June 1997 to August 2006.
7. For a thorough analysis of the World Bank Groups' role in shaping the country's mining governance, see Hatcher, *Regimes of Risk*.
8. World Bank, "Project Information Document."
9. See Campbell's contribution to this collection, as well as Campbell, *Regulating Mining*; Campbell, *Mining in Africa*; Hatcher, *Regimes of Risk*; and Gudynas, "The New Extractivism."
10. World Bank, "Striking a Better Balance," 10.
11. Ibid., 3. Additionally, any evaluation of the weight of the Bank over mining should encompass the private arms of the Group: the IFC, which holds a mining portfolio of US$500 million, and Multilateral Investment Guarantee Agency (MIGA), whose own portfolio currently stands at US$240 million. See MIGA, "Supporting Mining Investments." The IFC is a for-profit organisation which aims to support the growth of the private sector in developing countries. MIGA promotes foreign direct investment in developing countries. Its main activities are providing political risk insurance against certain non-commercial risks to investments.
12. Gagné-Ouellet, "Regulatory Framework."
13. Wu, "Mineral Industry of Mongolia," 15.1. (Our emphasis).
14. Connors, "Mining Law and Policy," 12; High, "The Cultural Logics of Illegality," 6; and Johnston, "Mongolia – Oyu Tolgoi," 2.
15. Packard and Khurebold, "The Changing Legal Environment."
16. High, "The Cultural Logics of Illegality," 6.
17. See Campbell's contribution to this collection.
18. Lahiri-Dutt and Dondov, "Informal mining in Mongolia," 5.

19. Ibid, 6.
20. Exact numbers are difficult to assess notably because of the wide diversity in nomadic herder households. For instance, while Lahiri-Dutt and Dondov agree on 213,000 (in 2014), Jacobs suggests a much higher number 800,000 (40% of the population). See Jacobs, "Winter Leaves Mongolian."
21. Urantsooj, President of the Center for Human Rights and Development, Personal Communication, 2012.
22. Mendoza et al., "Extractive Industries, Children," 4.
23. Pilling, "Mongolia Open to Talks."
24. As part of a larger US$4.4 billion loan package by multiple international lenders.
25. Langfitt, "Mineral-Rich Mongolia." Although it was expected to further accelerate to a staggering 20% in 2012 (Pilling, "Mongolia Open to Talks."), it fell quite short of expectations and decelerated to 12%. This was blamed on falling commodity prices and a slowdown in China, as well as internal factors, notably the changes in the country's investment regime in 2013 and a dispute between the government of Mongolia and Oyu Tolgoi's investors, a dispute which was to last for the following two years (2013–2015).
26. Mendoza et al., "Extractive Industries, Human Development," 4; and Ernst & Young, "Mongolia. Mining and Tax," 7.
27. Lahiri-Dutt and Dondov, "Informal mining in Mongolia," 4.
28. On the topic, see Blunt and Sainkhuu, "Serendipity and Stealth"; and Blunt, "Whose Resources are they anyway?"
29. ASM are activities difficult to define in the Mongolian context, as well as at the global level. ASM "means different things to different people," observes the SDC. SDC, "SDC Experiences with Formalization," 8. While in this contribution we choose to pool 'artisanal' and 'small-scale mining' together, it is imperative to note that the term ASM encompasses a wide diversity in terms of legal frameworks, scale, demography and labour organisations. The SDC broadly defines ASGM [Artisanal and small-scale gold mining] in relation to large-scale mining: 'ASGM is a spontaneous self-organizing social system, while industrial mining is planned and centrally coordinated. Artisanal miners engage in mining to earn a living, while industrial mining (large, medium and small-scale) is driven by profit expectations.' SDC, "SDC Experiences with Formalization," 8.
30. While ASM did exist prior to the socialist period, Grayson et al. argue that it completely disappeared during the Soviet rule. Grayson et al., "The People's Gold Rush," 2.
31. SDC, "Background."
32. World Bank, "Mongolia Poverty Assessment," 28.
33. Ibid., 28; Khongorzul, "Mongolia and Central Asia," 128. Before the transition, social inequalities were negligible and poverty was virtually inexistent in rural areas. IFAD, "Rural Poverty in Mongolia."
34. Informal mining began with the closing of the state-owned Nalaikh coal mine in 1995 when the urban-poor in Ulaanbaatar began to scavenge coal for the region's severe winters. Lahiri-Dutt and Dondov, "Informal mining in Mongolia," 4.
35. SDC, "Factsheet: Sustainable Artisanal Mining Project," 1.
36. It is estimated that in the aftermath of these *zuds* 15,000 herders had lost all their animals), notwithstanding the fact that all the herders who might not have lost their entire stock were severely impacted. Werger and Staalduinen, *Eurasian Steppes*, 518.
37. UNEP, "Analysis of Formalization Approaches," 4; SDC, "Responsible and Sustainable Artisanal Mining."
38. UNEP, "Analysis of Formalization Approaches," 4.
39. It is interesting to note that for High, ASM might not be entirely driven by poverty in the case of herders who undertake the activity. The author observed wealthy herders turning towards ASM as a means to increase a form of wealth which is 'less visible and measurable' than pastoral wealth, which 'is constantly visible and countable.' In doing so, herders avoid the social obligations which come attached to an increase in visible wealth. High, "Wealth and Envy," 9–10.
40. High observes that during the socialist era, mining companies had a low gold recovery rate due to outdated technical equipment and as such, these sites have provided 'a relatively rich

and readily accessible resource for artisanal and small-scale gold miners.' High, "Gold Mining in Mongolia," 1.

41. Stanway, "Mongolia's 'Ninja' Miners."
42. SDC, "Responsible and Sustainable Artisanal Mining," 2.
43. See Grayson et al., "The People's Gold Rush," 2. The term Ninja originates from the fact that miners carry green panning bowls on their backs, which resemble the turtles' shells of the main characters in the movie *Teenage Mutant Ninja Turtles,* which came out in 1990. Today, the term is widely used by all stakeholders in the sector. See High, "The Cultural Logics of Illegality," 9.
44. Villegas, "Ninja Miners and Rural." Placer means mineral concentration that has been separated from its first subsoil location as a result of erosion and mechanic processes and that are located as layers after reformation (see Art. 4.1.10 of Mongolia's 2006 Mining Law). In the case of gold, there is the 'left behind' by mining companies placer, and the placer in virgin deposits that is not in the reserves of mining companies. Grayson et al., "The People's Gold Rush," 7.
45. According to Lahiri-Dutt and Dondov, fluorspar would now be the second in importance in terms of ASM employment. "Informal mining in Mongolia," 4.
46. Seasonal ninjas work on average from late April until late September.
47. SDC, "Responsible and Sustainable Artisanal Mining," 1.
48. EITI and National Council of the Mongolia EITI, "Mongolia Extractive Industries," 30.
49. In a recent study of the sector, Isabel Cane et al. state that 64% of the ninjas considered their income from artisanal mining to be 20 times higher than other sources of income. Cane et al., *Responsible Mining in Mongolia*, 73.
50. Lahiri-Dutt and Dondov, "Informal mining in Mongolia," 4; and Grayson et al., "The People's Gold Rush," 2.
51. There are no definitions of ASM in the Mongolian Mining Law of 2006, nor in its 2006 revisions.
52. SDC, "Responsible and Sustainable Artisanal Mining," 1.
53. Heemskerk, "Gender Assessment of Small-Scale Mining," 7; SDC, "Responsible and Sustainable Artisanal Mining," 1; and Uitterdijk Appel, "Mongolia: From Zero," 1.
54. World Bank, "Project Information Document."
55. Lahiri-Dutt and Dondov, "Informal mining in Mongolia," 6.
56. SDC, "Responsible and Sustainable Artisanal Mining," 1.
57. High, "Wealth and Envy," 3. Also see High, "The Cultural Logics of Illegality."
58. See High, "Gold Mining in Mongolia"; and Purevjav, "Human Rights." While mercury use was made illegal in 2008 it is still widely available.
59. Purevjav, "Human Rights." Other factors contributing to the negative image of the sector are notably issues of child labour on ASM sites, tax evasion, rise in crime and additional pressure on local administrations. SDC, "The Small-Scale Mining," 7.
60. See Campbell's contribution to this collection.
61. Lahiri-Dutt and Dondov, "Informal mining in Mongolia," 11.
62. Ibid.
63. World Bank, "Mongolia. A Review," 22; and SDC, "SDC Experiences with Formalization," 16–17.
64. SDC, "SDC Experiences with Formalization," 16–17.
65. Ibid.
66. See Campbell's contribution to this collection.
67. Itty, "Responsible and Sustainable," 4.
68. Our translation. Personal Communications, August 2015.
69. Cited in Wacaster, "The Mineral Industry of Mongolia," 17.2.
70. Cited in Mendoza et al., "Extractive Industries, Children," 18; and Namkhaijantsan, "Learning-by-Doing," 60.
71. SDC, "Background."
72. UNEP, "Analysis of Formalization Approaches," 6; and SDC, "Background."
73. SDC (Swiss Agency for Development and Cooperation) employee, Personal Communications, October 2013.
74. Hall et al., *Powers of Exclusion*.

75. Hall, Hirsch and Li's (Ibid., 5.) further explain that 'regulation' is often but not exclusively associated with the state and legal instruments, sets of rules regarding access to land and conditions of use; 'force' excludes by violence or the threat of violence; 'the market' is a power of exclusion as it limits access though price and through the creation of incentives to lay more individualised claims to land; and 'legitimation' establishes the moral basis for exclusive claims, and indeed for entrenching regulation, the market and force as politically and socially acceptable bases for exclusion.
76. SDC, "SDC Experiences with Formalization," 16–17.
77. A 'ger' is a traditional tent in which many Mongolians live.
78. World Bank, "Mongolia. A Review," 22–3.
79. Hall et al., *Powers of Exclusion*.
80. SDC, "The Small-Scale Mining," 22–3. A notable exception would be *Soum* (district) governors, who, in light of their direct knowledge of the importance of the sector in terms of local livelihood, have traditionally been supportive of ASM (SDC employee, Personal Communications, October 2013).
81. UNEP, "Analysis of Formalization Approaches," 6. Notably via projects by the Mining, Minerals and Sustainable Development project and a World Bank's Communities and Small-scale Mining initiative.
82. Up to 34% where non-State funding sources are used to determine proven reserves or up to 50% ownership where State funds are used (Art. 5.4).
83. World Bank, "Project Information Document."
84. The Interim Regulation No.33 on Gold Extraction by Non-Industrial Means (IR33), which had a 12 months life span, allowed for artisanal gold mining to take place on already licenced land. *Soum* (district) governments were given the authority to conclude agreements with mining license holders, allowing individual miners to rework tailings and waste rock. In turn, the mined gold was to be sold to the licence holder. IR33 had a low participation rate. This was mainly due to the trying need to secure a tripartite agreement between the mining company, the ninjas and the local administration; and due as well to bureaucratic requirements. Grayson et al., "The People's Gold Rush," 67–8. Barely a year after its inception, IR33 was replaced by a slightly modified version of the regulation (IR53). While the latter was also short lived, some local officials continued to implement the regulation (Ibid., 60.).
85. SDC, "The Small-Scale Mining," 8.
86. The legal framework built on several consultations which were undertaken in 2009–2010 with more than 400 stakeholders, including members of parliament, central and local government officials, ASM representatives, NGOs and artisanal miners. UNEP, "Analysis of Formalization Approaches," 8.
87. MASM, *Small-Scale Mining*.
88. The new Law was met with the amendment of the Mineral Law, the Land Law and the Personal Income Tax Law.
89. SDC, "The Small-Scale Mining," 22.
90. Itty, "Responsible and Sustainable," 3.
91. SDC, "An Update of ASM."
92. UNEP, "Analysis of Formalization Approaches," 8.
93. Itty, "Responsible and Sustainable"; and SDC employee, Personal Communications.
94. Itty, "Responsible and Sustainable," 3. The core objective of the project is "to develop an economically sustainable, environmentally responsible and human rights-based ASM sector in Mongolia benefiting from, and contributing to, global best practice regarding ASM." SDC, "Factsheet: Sustainable Artisanal Mining Project," 1.
95. SDC. "SDC Experiences with Formalization," 25.
96. MASM, *Small-Scale Mining*.
97. Itty, "Responsible and Sustainable," 4. According to the SDC's 2015 Annual Report, 8044 miners (18% of which were women) had been trained in occupational health and safety by the MoL and the National Emergency Management Agency, as well as the support of the SDC. SDC. "Annual Report," 20.

98. SDC employee, Personal Communications.
99. SDC, "Annual Report," 20.
100. SDC, "Responsible and Sustainable Artisanal Mining," 1. Furthermore, the SDC has assisted communities in obtaining the certification "Fairmined Eco Gold," the first in Asia to obtain such certification. The project developed mercury-free gold-processing plants. See SDC, "Factsheet: Sustainable Artisanal Mining Project," 2.
101. A province in Mongolia.
102. Our translation. Personal Communications, August 2015.
103. Ibid.
104. By 2012, the sector was contributing between US$9 and 10 million to the local economy, and 821 miners were contributing 82 million MNT in income tax. SDC, "Project Monitoring Sheet," 1. Also see SDC, "An Update of ASM."
105. SDC, "An Update of ASM."
106. Ibid.
107. SDC, "Annual Report 2015."
108. SDC, "An Update of ASM."
109. Cited in High, "Gold Mining in Mongolia," 2.
110. ARM and SCS Global Services, "Artisanal Miners Now"; ARM and SCS Global Services, "Ecological Gold from Mongolia."
111. Several issues with the 2010 regulation are being discussed by parliamentarians including: attempts to scale up artisanal mining into small-scale mining through miners partnerships and how many miners these should include; underground mining; whether basic deposits should be the object of ASM; the amount of land that should be allocated for ASM; the use of explosives; restrictions on equipment capacity; allocation of responsibility for occupational health and safety; environmental rehabilitation measures and costs and the use of mercury; depth of digging. SDC, "SDC Experiences with Formalization," 24. Furthermore, the new legislation has failed to tackle the practice whereby legal licence owners are subcontracting their plot to illegal miners, and in so doing, exposing the latter to serious human rights concerns, notably a weak safety environment.
112. See Campbell's contribution to this collection.

Bibliography

ARM (Alliance for Responsible Mining), and SCS Global Services. "Artisanal Miners Now Can Prove They Are Legal." *MINING.com*, May 27, 2014. http://www.mining.com/web/artisanal-miners-now-can-prove-they-are-legal/.

ARM, and SCS Global Services. "Ecological Gold from Mongolia: Artisanal Miners Obtain Fairmined Certification." *MINING.com*, February 27, 2015. http://www.mining.com/web/ecological-gold-mongolia-artisanal-miners-obtain-fairmined-certification/.

Blunt, P. "Whose Resources Are They Anyway? 'Development Assistance' and Community Development Agreements in the Mongolian Mining Sector." *Progress in Development Studies* 14, no. 4 (2014): 383–99.

Blunt, P., and G. Sainkhuu. "Serendipity and Stealth, Resistance and Retribution: Policy Development in the Mongolian Mining Sector." *Progress in Development Studies* 15, no. 4 (2015): 371–85.

Campbell, B., ed. *Regulating Mining in Africa: For Whose Benefit?*. Uppsala: Nordiska Afrikainstitutet, 2004.

Campbell, B., ed. *Mining in Africa. Regulation and Development*. London, Ottawa and Uppsala: Pluto, IDRC and Nordica Africa Institute, 2009. http://www.idrc.ca/EN/Resources/Publications/openebooks/450-5/index.html.

Cane, I., A. Schleger, S. Ali, D. Kemp, N. McIntyre, A. Lechner, P. McKenna, B. Dalaibuyan, Nena Bulovic, and K. Lahiri-Dutt. *Responsible Mining in Mongolia: Enhancing Positive Engagement*. St-Lucia: Sustainable Minerals Institute, The University of Queensland, 2015. http://espace.library.uq.edu.au/view/UQ:362422.

Connors, J. "Mining Law and Policy Developments in Mongolia: Considerations for Foreign Investors." *Baker & McKenzie International*, February 10, 2011. http://www.bakermckenzie.com/files/Publication/6c2c6e96-263f-4ff1-82d8-329718c366a7/Presentation/Publication

Attachment/24c777ae-04a0-4880-9caf-d391b0ae7995/pn_tokyouraniummining_mongolia_feb10.pdf.

EITI (Extractive Industries Tranparency Initiative), and National Council of the Mongolia EITI. "Mongolia Extractive Industries Transparency Initiative (MEITI). Mongolia Eight EITI Reconciliation Report 2013." *Extractive Industries Transparency Initiative*, December 15, 2014.

Ernst & Young. "Mongolia. Mining and Tax Guide 2012/13." In *The 2012 Worldwide Corporate Tax Guide*, 783–91, 2012. http://www.ey.com/Publication/vwLUAssets/Worldwide_corporate_tax_guide_2012/$FILE/WCTG_2012_Worldwide_Corporate_Tax_Guide.pdf.

Gagné-Ouellet, S. "Regulatory Framework Revision and Mining Regime Reform in Mali: Degrees of Rupture and Continuity." In *Modes of Governance and Revenue Flows in African Mining*, edited by B. Campbell, 47–100. International Political Economy Series. Palgrave Macmillan, 2013.

Grayson, R., T. Delgertsoo, W. Murray, M. Tumenbayar, U. Tuul, D. Bayarbat, and C.-E. Baatar. "The People's Gold Rush in Mongolia. The Rise of the 'Ninja' Phenomenon." *World Placer Journal* 4, no. Special Issue (2004): 1–112.

Gudynas, G. "The New Extractivism of the 21st Century: Ten Urgent Theses about Extractivism in Relation to Current South American Progressivism." Americas Program Report. Washington, DC: Center for International Policy, January 21, 2010.

Hall, D., P. Hirsch, and T. M. Li. *Powers of Exclusion: Land Dilemmas in Southeast Asia*. Honolulu: University of Hawai'i Press, 2011. http://www.uhpress.hawaii.edu/p-7556-9780824836030.aspx.

Hatcher, P. *Regimes of Risk*. International Relations & Development Collection. London: Palgrave Macmillan, 2014. http://www.palgraveconnect.com/doifinder/10.1057/9781137031327.

Heemskerk, M. "Gender Assessment of Small-Scale Mining in Mongolia." *The Asia Foundation*, October 2013.

High, M. "Wealth and Envy in the Mongolian Gold Mines." *Cambridge Anthropology* 27, no. 3 (2008): 1–18.

High, M. "Gold Mining in Mongolia." Chap. 221 in *Encyclopaedia of the History of Science, Technology, and Medicine in Non-Western Cultures.* Dodrecht: Springer Science & Business Media, 2014.

High, M. "The Cultural Logics of Illegality: Living outside the Law in the Mongolian Gold Mines." In *Change in Democratic Mongolia: Social Relations, Health, Mobile Pastoralism, and Mining*, edited by J. Dierkes, 249–70, Pre-copyedited version. Leiden: Brill Publishers, 2012. http://www.research.ed.ac.uk/portal/files/12353426/Cultural_Logics_of_Illegality.pdf.

IFAD (International Fund for Agricultural Poverty). "Rural Poverty in Mongolia." IFAD. *Rural Poverty Portal*. Accessed November 22, 2013. http://www.ruralpovertyportal.org/country/home/tags/mongolia.

Itty, P. "Responsible and Sustainable Artisanal Mining. A Golden Opportunity for Mongolia." *Asia Brief. East Asia Division*, March 4, 2013.

Jacobs, A. "Winter Leaves Mongolians a Harvest of Carcasses." *The New York Times*, May 19, 2010. http://www.nytimes.com/2010/05/20/world/asia/20mongolia.html.

Johnston, L. "Mongolia – Oyu Tolgoi Copper/Gold/Silver Mine Project Trip Report." Trip Report. Washington, DC: USAID (United States Agency for International Development), June 2011. http://mongolia.usaid.gov/wp-content/uploads/Mongolia-Oyu-Tolgoi-Trip-Report.pdf.

Khongorzul, D. "Mongolia and Central Asia in Transition." *Ritsumeikan Journal of Asia Pacific Studies* 26 (2009): 117–48.

Lahiri-Dutt, K., and H. Dondov. "Informal Mining in Mongolia: Livelihood Change and Continuity in the Rangelands." *Local Environment* 13 (May 2016): 1469–6711. doi:10.1080/13549839.2016.1176012.

Langfitt, F. "Mineral-Rich Mongolia Rapidly Becoming 'Mine-Golia'." Special Series Mongolia Booms. *National Public Radio*, May 23, 2012. http://www.npr.org/2012/05/21/152683549/mineral-rich-mongolia-rapidly-becoming-minegolia.

MASM (Mongolian Artisanal Miners' United Umbrella Association NGO). *Small-Scale Mining and Human Rights Situation of Artisanal Small-Scale Miners in Mongolia*. Geneva: Universal Periodic Review, 2014. https://uprdoc.ohchr.org/uprweb/downloadfile.aspx?filename=1661&file=EnglishTranslation.

Mendoza, R. U., H. MacArthur, and A. Ong Lopez. "Extractive Industries, Children and Inclusive Growth: An Analysis of Mongolia, Papua New Guinea and Timor-Leste." Working Paper. UNICEF, January 2012. http://share.pdfonline.com/2c004f9c7d4340128a38c861e8da6724/UNICEF%20EXTRACTIVES%20PAPER%20JAN2012.pdf.

Mendoza, R. U., H. MacArthur, and A. Ong Lopez. "Extractive Industries, Human Development and Child Rights." Ulan Bator, October 20–22, 2011. http://www.undp.mn/mining/papers/MONGOLIA_CONFERENCE_mendoza.pdf.

Naito, K., H. Myoi, J. Otto, D. Smith, and M. Kamitani. "Mineral Projects in Asian Countries – Geology, Regulation, Fiscal Regimes and the Environment." *Resources Policy* 24, no. 2 (1998): 87–93. doi:10.1016/S0301-4207(98)00012-9.

Namkhaijantsan, D. "Learning-by-Doing: Civil Society Engagement in Mining in Mongolia." In *Advancing the EITI in the Mining Sector. a Consulation with Stakeholders*, edited by P. Mitchell, F. Paris, and C. Eads, 59–63. EITI (Extractive Industries Transparency Initiative), 2009. https://eiti.org/files/MINING%20Compressed.pdf.

Packard, C., and O. Khurebold. "The Changing Legal Environment for Mining in Mongolia." *Anderson & Anderson LLP*, 2010. http://zotero://attachment/220/.

Pilling, D. "Mongolia Open to Talks on Investment Law." *Financial Times*. May 29, 2012. http://www.ft.com/cms/s/0/3215c23a-a98e-11e1-9972-00144feabdc0.html.

Purevjav, B. "Human Rights and Mongolia's Small-Scale Mining Sector." October 29, 2014. http://asiafoundation.org/in-asia/2014/10/29/human-rights-and-mongolias-small-scale-mining-sector/.

SDC. "SDC Experiences with Formalization and Responsible Environmental Practices in Artisanal and Small-Scale Gold Mining in Latin America and Asia (Mongolia)," 2011. https://www.eda.admin.ch/content/dam/deza/en/documents/publikationen/Diverses/216063-artisanal-gold-mining_EN.pdf.

SDC. "The Small-Scale Mining Legal Framework Development Process." *Sustainable Artisanal Mining Project*, 2011. http://www.sam.mn/images/documents/erh_zui.pdf.

SDC. "Project Monitoring Sheet," 2013.

SDC. "Background." Sustainable Artisanal Mining Project, 2015. http://www.sam.mn/info_en.php?url=intro.

SDC. "An Update of ASM in Mongolia 2015." Knowledge Hub Desk, Sustainable Artisanal Mining Project, February 18, 2016. http://www.sam.mn/news_en.php?title=An-update-of-ASM-in-Mongolia-2015

SDC. "Annual Report 2015." Swiss Agency for Development and Cooperation (SDC) in Mongolia, Ulaanbaatar. https://www.eda.admin.ch/content/dam/countries/countries-content/mongolia/en/SDC-in-Mongolia-Annual-Report-2015_EN.pdf.

SDC. "Factsheet: Sustainable Artisanal Mining Project." Sustainable Artisanal Mining Project, December 2015. www.swiss-cooperation.admin.ch/mongolia.

SDC. "Responsible and Sustainable Artisanal Mining – A Golden Opportunity for Mongolia." *Asia Brief. East Asia Division*, March 2013.

Stanway, D. "Mongolia's 'Ninja' Miners Help Sate China Lust for Gold." *Reuters*, April 19, 2012, U.S. Edition edition. http://www.reuters.com/article/2012/04/19/us-mongolia-gold-idUSBRE83I08V20120419.

Uitterdijk Appel, P. W. "Mongolia: From Zero to 100 000 Small Scale Miners in Less than 10 Years." Manilla, 2005. http://asmasiapacific.org/wp-content/uploads/2014/07/Appel2005_Mongolia-From0To1000-1.pdf.

UNEP (United Nations Environment Programme). "Analysis of Formalization Approaches in the Artisanal and Small-Scale Gold Mining Sector Based on Experiences in Ecuador, Mongolia, Peru. Tanzania and Uganda. Mongolia Case Study," June 2012.

Villegas, C. "Ninja Miners and Rural Change in Mongolia." Blog. *Artisanal Small Scale Mining in and around Protected Areas and Critical Ecosystems*, August 26, 2013. http://www.asm-pace.org/blog/item/11-ninja-miners-and-rural-change-in-mongolia.html.

Wacaster, S. "The Mineral Industry of Mongolia." 2012 Minerals Yearbook. Mongolia (Advance Release). U.S. Geological Survey Mineral Yearbook 2012. U.S. Department of Interior and U.S. Geological Survey, June 2014.

Werger, M., and M. Staalduinen. *Eurasian Steppes. Ecological Problems and Livelihoods in a Changing World*. Dodrecht: Springer Science & Business Media, 2012.

World Bank. "Mongolia Poverty Assessment in a Transition Economy." Rural and Social Development Operations Division, China and Mongolia Department, East Asia and Pacific Regional Office, June 27, 1996. http://www-wds.worldbank.org/servlet/WDSContentServer/WDSP/IB/1996/06/27/000009265_3961214164741/Rendered/PDF/multi0page.pdf.

World Bank. "Striking a Better Balance: Volume 1. the World Bank Group and Extractive Industries." Final Report. Extractive Industries Review (EIR). Jakarta: World Bank Group, December 2003. https://openknowledge.worldbank.org/handle/10986/17705.

World Bank. "Mongolia. A Review of Environmental and Social Impacts in the Mining Sector." Discussion Papers. Environment and Social Development. East Asia and Pacific Region, May 2006.

World Bank. "Project Information Document: Concept Stage." Mongolia Mining Sector Technical Assistance Project. Washington, DC: Oil, Gas and Mining Policy Department, 2008.

World Bank. "Mongolia: Overview." World Bank Website, 2015. http://www.worldbank.org/en/country/mongolia/overview.

Wu, J. C. "The Mineral Industry of Mongolia." 1999 Minerals Yearbook Mongolia. U.S. Geological Survey Mineral Yearbook 1999. U.S. Department of Interior and U.S. Geological Survey, 1999.

Property rights and large-scale mining: overlapping claims at and around mining sites at the Democratic Republic of Congo and Zambia

Sarah Katz-Lavigne

ABSTRACT
Large-scale mining in the Democratic Republic of Congo and Zambia, characterised by limited statehood, does not fit neatly into classical property rights theory. Institutional change is messy and unlikely to be as simple as a government setting aside a concession for a mining company. The persistence of contestation around mining sites is a sign that overlapping claims to rights over resources, the emergence and reconfiguration of coalitions around property rights enforcement, and resistance and conflict are likely to persist, companies' efforts to implement corporate social responsibility notwithstanding. The theoretical discussion is complemented by empirical evidence from the DRC and Zambia.

Introduction

Property rights (PR) underpin and structure economic life: by 'defining the parameters for the use of scarce resources and assigning the associated rewards and costs, the prevailing system of property rights establishes incentives and time horizons for investment, production, and exchange.'[1] A right is a claim that can be enforced through law, customs or conventions. A claim is a right with respect to another person, who has a corresponding duty towards the holder of the right. Interests serve as a basis through which a claim is linked to a right.[2] PR do not imply full and exclusive control over a resource (ownership); they can be better conceptualised as a bundle or web of interests.[3] From land to minerals, formalisation of rights has implications for these interests.

The persistence of plural, overlapping and contested PR regimes globally highlights the limitations of classical PR theory. A modified approach that explores how PR are enforced in contexts of overlapping claims, where states are unable and/or unwilling to do so, can be fruitfully applied to the large-scale mining (LSM) sector in the Democratic Republic of Congo (DRC) and Zambia, which exhibit limited statehood[4]: 'a state monopoly on the use of force and authoritative decision-making competence on the side of the state are either non-existent or only partially existent'.[5] In the DRC and Zambia the state has granted PR to LSM companies, even when it cannot enforce these claims and the local population considers them illegitimate. Overlapping, unresolved claims may lead to persistent conflict even as it influences and reconfigures the coalitions of interests that align to seek rents from what has been, for some, a profitable sector.

Rights definition and enforcement

Classical PR theory argues the economic efficiency of private property, compared to resources that are open for access or communally managed. This theory, developed for Europe and North America, predicts that as resource values rise, the benefits to private ownership will eventually be greater than the costs, and private PR will emerge.[6] Rational actors will see that factor prices are changing or new technology is emerging, and will want to capture the benefits.[7] The institutionalisation of private property has been linked to variations in prosperity across the world's regions.[8] In development theory, Acemoglu and Robinson have argued that the evolution of favourable economic institutions, namely a relatively equitable allocation of PR, provided the impetus for investment and innovation in countries like the US.[9]

PR include use rights (the right to access, withdraw from, or exploit a resource) and rights to control or make decisions about a resource (management, exclusion, and alienation – by renting, selling, or giving away – of rights). In formal systems of law and classical PR theory, the emphasis has been on the ability to transfer rights. Around the world, many societies view resources like land as held by a group rather than any individual, and therefore more costly and difficult to transfer. Even in legal systems with high formalisation, bundles of rights may not be vested in the same person.[10] Mining encompasses at least two sub-bundles: surface land and subsoil resources.

PR may be granted, but will not hold much sway if they cannot be enforced. There are several strategies through which parties can enforce their rights against encroaching groups. First-party measures involve unilateral use of force. If first-party enforcement is not possible or is against the law, second-party measures involve working privately with other claimants to implement an arrangement.[11] Third-party measures refer to recourse to intervention by an outside actor – generally the state.[12] The mix of strategies is context-specific and will depend on the extent to which land is scarce[13] (the demand side), but also on the supply of rights, including state capacity or the capacity and legitimacy of customary authorities. Where the state does not define and enforce PR; where it defines but does not enforce them; or where it selectively enforces them, the protection of property in a context of rising scarcity is likely to involve reliance on first (self-enforcement, e.g. private security), second (interest-based coalitions) and non-state third party (such as an armed group) measures.[14] The 'ability to retain de facto rights in the face of someone else holding the *de jure* rights rests on political, wealth and violence advantages for the first possessor'.[15] Compared to other domestic claimants like local communities and those in artisanal and small-scale mining (ASM), companies possess vast capabilities in politics, wealth and violence, restrained only by factors like the need to be perceived as good corporate citizens.

The provision of well-defined, enforced PR is typically associated with Western statehood. Governments may have an incentive to supply rights to capture political and economic benefits, but PR may not emerge if the economic rents are not sufficient for governments to assume the costs. It is not only the demand for PR definition and enforcement that matters, but also the extent to which governments have incentives to supply them.[16]

Classical PR theory was developed in a specific socio-economic context and can't be assumed to apply elsewhere. In much of the global South, the state cannot or will not enforce rights to private property in a way typically associated with the Weberian state. This is why the DRC and Zambia have been characterised as areas of limited statehood.[17] Scholars have challenged the concept of the Weberian state in the African context, noting that the colonial state was fundamentally extractive[18] and distinguished between citizens and subjects.[19] However, even in the DRC – often used as an example of extremely limited statehood – the state may have a limited physical presence, but does possess political and economic reach. This is the case when it comes to lending authority to firms' PR. Even when the state cannot enforce these rights, they have political effects, including allowing companies to claim their actions are legitimate even if they violate national legislation. In the DRC, Chinese companies MKM and HUACHIN have violated Congolese legal measures on employment, the environment, contributions to community development, and more – all with government authority tolerance or collusion.[20]

Whatever the cause(s), contestation and conflict over resources endure in societies 'characterized by normative and legal pluralism such as post-colonial and post-socialist countries'.[21] In the DRC and Zambia, systems for organising property are faced with 'widespread legal uncertainty [and] resource conflicts'.[22] Customary social norms that previously allowed for the rational management of resources in common have become degraded, often as a result of intervention by a state without a high degree of legitimacy. In most instances, local norms and institutional arrangements persist[23]; multiple actors lay claim to resources, but are unable to exclude other parties. This may be at the root of long-lasting conflict as property rights regimes transform into contested access.[24] The principles of property rights theory can be helpfully applied in making sense of these overlapping claims and how they are enforced in contexts where the state does not step in to do so.

Land and property rights

Land issues have received significant attention in countries where much of the population is engaged in agriculture. There is extensive literature on forms of tenure, conflict over land, and the efficiency and distributional implications of the privatisation of communally held lands. Much literature has focused on 'identifying factors that motivate the transformation of property rights toward increased privatization, or in explicating the desirable outcomes of such transitions.'[25] Mwangi noted that: 'There is less understanding of how the solution of individual property is reached or what makes the new property assignment binding'.[26] Her study of the subdivision of communal Maasai land in Kenya analyses the distributional outcomes and why many Maasai who initially favoured the process were disappointed:

> … the process by which property rights change is so full of politics, that to better understand this process of change, we need to move beyond economic models of relative price changes and state enforcement, to models that accommodate competition by actors and the possibility that state actors may not necessarily provide the arbitration/enforcement that is often taken for granted.[27]

Mwangi found that the efficiency gains that classical property rights theory predicts will result from private or individual title may be much smaller than expected due to persistent competition and conflict. The form conflict takes, and how intense it becomes, is a function of the degree of concentration of benefits in the new dispensation, and the diversity of interests among parties bargaining for land. There is not always an easy recipe or clear formulation for the distribution of benefits, so the answer to how these issues are resolved is a contextual one. Agencies of the state may not intervene, or may be involved but not impartial. The results of bargaining are dependent not only on actors' influence and power, but also the extent to which the new distribution is congruent with institutional norms in place. If the proposed new structure is highly concentrated, institutional change may be delayed or not occur. In this case, certain actors may have recourse to the state's authority to put an end to the conflict. Alternatively, actors with more resources and a stronger position may be able to impose their will on weaker parties.[28] Similar lessons can be applied to countries in which LSM investors are granted PR that overlap with existing claims.

Yet even weaker parties may have some influence, potentially resulting in recurrent conflict.[29] Violence has high costs in human and material terms. Weaker actors' decision to engage in actions putting them at risk of violent expulsion – such as encroaching on an LSM concession – suggests that a cost-benefit analysis leads them to view the possibility of violent removal as a least worst alternative.

Informal institutions

Institutions are the societal rules that govern how people interact with each other; these can be either formal or informal constraints on human behaviour. Institutions – not to be confused with organisations – help lower uncertainty by structuring people's daily lives. Institutions consequently shape the economic, political and social incentives that impact human exchange.[30]

In much of Africa, the formal legal system is not the only – nor, in some cases, the main – institution responsible for the allocation of property rights. In a number of countries PR to land have long been allocated through informal institutions such as customary authority and gender. In the DRC (a country emerging from conflict) and Zambia, where investors' interest in mineral and fuel resources has grown, the national legal system is being increasingly used to allocate PR, particularly to foreign investors. These dynamics make it important to ascertain the extent to which informal institutions for land allocation have been displaced, and assess the distributional and conflict-related implications.

While ASM has been presented as poorly organised and chaotic[31] due to the absence of robust, formal rights to property, artisanal miners may be operating on the basis of customary title to resources. Even if operating illegally, they may view their customary title to resources as more durable and legitimate.[32] In the DRC, artisanal gold mining is highly structured and locally adapted.[33] Implementing a formal system is likely to upend 'existing dependencies, reciprocal relations and power balances',[34] with consequences that may be unforeseen by policy-makers. LSM investment does not take place in a vacuum and is likely to disrupt and reconfigure property rights arrangements. The consequences may not be all bad. A case study of the Chinese mining company Huachin in Katanga, DRC described how certain miners could do better extracting minerals from LSM concessions, and selling directly to Huachin, than within the conditions set by an ASM cooperative.[35] Such examples are likely to be the exception, but they highlight the importance of a contextual understanding.

Gender is another important, often overlooked, institution that impacts how property rights are distributed. Agarwal argued that 'gendering the land question remains critical'.[36] This is likely to be the case for LSM and land usage; more research is needed to ascertain the distributional impacts of increased investment on women's land rights and capital accumulation. There is growing evidence that gendered patterns of land ownership and access, and of the costs and benefits of economic activity, need to be taken into account. In September 2015 ActionAid Zambia published a report describing a range of negative environmental, sociocultural, economic and political impacts of the mining industry on women, including loss of access to land and forest resources. Compensation is clearly gendered:

> Compensation packages did not take into account the loss of access to common forest and water resources that women have depended on for food, medicines and income for many generations. Compensation also tended to go to men and be based on the losses that men had incurred. Without women's losses being factored in and with no access or control over the money, women were thus doubly disadvantaged [...] Political decisions concerning relocation and compensation rarely captures the needs, issues and concerns of women. This is partly because consultation processes do not provide for effective participation of women. The consequence of this is that decisions made by the community tend to favour men more than women including where to build infrastructure and the types of community programmes.[37]

LSM in the DRC and Zambia

In human rights reporting and the political economy literature on mining, much of the focus has been on the dispossession of artisanal miners or communities and on conflict – at times violent – between investors and artisanal miners,[38] or companies and communities. For artisanal miners, losing access to minerals is analogous to the loss of farmland by those who depend on agriculture – particularly in areas where, due to displacement; armed conflict; the desire for a higher standard of living; and/or the inability to generate enough income through farming, agriculture is not a viable option. Artisanal miners do not always originate from the places where they have settled to extract resources. Diggers from other areas will be less integrated into local customary arrangements and may have a weaker claim to resources.[39]

Theorising property and the distributional impacts of enclosing tracts of land is therefore crucial when it comes to LSM. There are distributional consequences to enclosing large areas previously mined, inhabited and/or cultivated by local inhabitants, including the configuration and reconfiguration of coalitions of actors in the different phases of a mining project.[40] Outbreaks of violent conflict at mining sites, such as violent repression by security forces or police at companies' request, point to the lack of legitimacy of a given property rights regime in the eyes of a weaker group like artisanal miners, leading them to seek access, even once access has been criminalised, to mining sites. There is a need to examine the institutional economy underpinning mining activities in LSM areas, specifically the coalitional arrangements that govern a particular mining site, including miners' and communities' continued access to resources. Outbreaks of violence may be a result of deliberate enforcement strategy or a sign of the breakdown or reconfiguration of a coalition of actors governing resource access. Rather than skipping over the messy middle, there is a need to examine the actors involved and the extent to which they influence or mediate processes of enclosure.[41] For example, Geenen interrogated proposals by policy-makers to formalise artisanal mining in the DRC, particularly the assumption that granting artisanal miners full legal title, which they could transfer, would solve the economic and social concerns linked to ASM.[42]

The application of a modified property rights framework to LSM in the DRC and Zambia sheds light on different rights to which actors make claims, often claims that overlap. Human rights organisations have called attention to the dispossession of local communities by mining firms, depriving them of land, housing, clean water, and more.[43] New mining investment therefore affects overlapping claims to resources above ground and subsoil. If rights to these resources have not been assigned, or if they have been assigned but are not recognised for mitigation or compensation measures, there is potential for conflict.

The DRC and Zambia are appropriate case studies to illustrate and analyse these dynamics. They share a mineral-rich geographical area, the Copperbelt. Both countries have a long history of LSM going back to colonial times, a record of state ownership of mining, and a recent push for privatisation, deregulation and attracting new foreign investors. Both governments have sought to maximise the rents obtained from the mining sector, including for personal benefit, though misappropriation has been particularly serious in the DRC. Large numbers of people are engaged in ASM and subsistence agriculture in both states. Both countries exhibit limited statehood, though unlike the DRC Zambia has remained peaceful since independence, albeit with recent years characterised by unstable democracy and policy uncertainty in the mining sector. Both countries have experienced conflict over land and resources, pitting mining companies against artisanal miners and local communities.

Legal frameworks governing mineral extraction: not just about the minerals

In Africa the national legal framework is an important mechanism for the allocation of mining rights. Mining law includes the

> definition of minerals, ownership of resources, law relating to the right to mine, conditions of governing [*sic*] the issue and holding of mining rights to enable private parties [*sic*] access mineral resources in a country, and the relationship between mineral and surface right holders.[44]

In Zambia, as in many countries including Kenya and Botswana, 'the State holds the property rights to all minerals [...] on behalf of the people'[45] (though in practice, the state recognises other parties' right to use the resources). This is the case regardless of other rights to the land or property. This legal provision makes it possible for the Zambian state to provide rights to minerals on private land as an incentive for investors, rendering it unnecessary for the holder of the mining right to purchase the land.[46] The rights to the minerals and the land are two separate things.

Another approach is the lease system (also known as the 'regalien' system) common in the Global South: 'the state owns the mineral title and the miner derives his or her right to extract minerals from a tenure granted by the state, as opposed to one granted by the land owner.'[47] Once a mining firm receives tenure to a mineral, the company is considered to own the mine and ore body. This rarely resembles fee simple tenure, but rather several different rights and obligations bundled together; these obligations vary from one country to the next.[48] A major difference between the regalien approach and the system in Zambia is that in Zambia, the state cannot definitively alienate (sell) its minerals[49] but only grant interests to them. This is not the same as granting legal title to minerals or land. The state is the owner of the mineral and land titles. After extraction, land use rights as well as title to the minerals that have been extracted become vested in the mining licence holder.[50] The regalien system allows for mineral rights to be sold.[51]

There is also a distinction between different types of mining rights. The Zambian Mines and Minerals Development Act contains provisions for prospecting and mining licences, with specific requirements for each.[52] The Act also enables Zambian nationals to obtain a licence for small-scale mining.[53]

Zambian legislation clarifies the relationship between surface rights and mining rights. When someone is granted a mining right, it does not mean they automatically receive surface rights to the land. The state retains the rights to the surface. Surface title is key for holders of mining rights and surface owners. Mining rights holders are likely to have a variety of uses for the surface, while the owner of the land may continue to rely on it for agriculture or other activities.[54]

In Zambia the Mines and Minerals Development Act provides considerable rights to the holder of the mining right[55]: 'A mining right holder's right to use the surface is superior to that of the surface owner, i.e. the mining right is the *dominant interest.*'[56] Mining rights are granted irrespective of how the land is being used.[57] A holder of a mining licence can purchase the area where mining will take place, or lease the land from the surface owner. If the landowner does not wish to cooperate with the holder of the mining right, the state can step in and acquire the land. Before the state can obtain the land in this way, the mining right holder has to demonstrate that he or she has done everything possible to secure the land, or the right to it, in a peaceful manner.[58]

In the DRC, the Mining Code implicitly encourages expropriation of land in that mining rights are not linked to land rights.[59] This is not necessarily a problem, except in so far as a lack of clarity contains the potential for conflict. The Code does not fully define rights such as those to 'a body of water, fruit trees, cultural objects or community links, among others'.[60] In the absence of any reference to these types of rights and the amount of compensation to which claimants would be entitled, overlapping claims to resources are likely to persist, representing a potential risk of conflict.

LSM, distributional issues, and conflict

Distributional concerns underpin any given property rights arrangement.[61] In the DRC and Zambia, conflict between LSM companies and ASM miners, between companies and community members, and potentially between communities and ASM[62] has remained persistent despite multinational and state-owned enterprises' efforts to promote corporate social responsibility (CSR) to forestall conflict. No matter how it is defined, what is fundamentally at stake when companies implement CSR is not providing communities with a share of the benefits from mining, but firms' ability to secure and enforce their PR. Despite the extensive rhetoric, standards, academic analysis, and organisational infrastructure that have sprung up around CSR to fine-tune and refine these measures, CSR is best conceptualised as a tool among others for MNCs to enforce PR and ensure uninterrupted access to mining concessions.

The Congolese and Zambian governments have readily granted mineral rights to LSM investors,[63] even in the face of long-standing claims. Companies rely on government legitimisation of their claims, even when central governments do not ensure enforcement. Artisanal miners, communities and companies often claim the same resources, with a significant power differential between the parties.[64] Understanding the varying endowments of competing groups is important to infer how these endowments will impact the political and property rights outcome that emerges.[65]

Conflict over the distribution of a resource may arise in contexts in which the state redistributes property rights and compensation is not provided, or is not sufficient to satisfy all parties. It has proven difficult for different claimants to reach agreement on how much compensation should be paid and to whom, and who should provide it.[66] In Katanga, compensation for households evicted from the Luisha mining site on behalf of Chinese company Congo International Mining Corporation was set after an evaluation by the local *Chef de poste*. The individuals removed had been told they did not hold rights to the land where their homes stood, despite payments made to the local chief as per customary land usage. Members of the community did not know how compensation of 100–300 USD was arrived at, nor could they challenge the amount.[67]

Mining legislation has a role in clarifying property rights and responsibilities to avoid conflict. In Zambia, the 'mining right holder's right to use the surface and the circumstances under which the owner's rights are protected are strictly regulated under the mining laws.'[68] Before an individual can exercise his or her mining right, they must give notice to allow the owner of the surface to gather crops and move livestock.[69] Zambia's mining legislation contains limitations to dominant interest: 'Section 73 provides that the holder of a mining right shall exercise his or her rights reasonably and in a manner which minimally affects the interests of any owner or occupier of the land'[70] while 'section 76 imposes liability for damage in certain situations where holder's activity was not even negligent'.[71] Yet while the Mines Act allows the government to terminate a right to mine due to a breach of statutory obligations, the government has never done so.[72]

Like Zambia, the DRC receives significant investment in copper and cobalt in the former Katanga Province, which in August 2015 was divided into four provinces according to a provision of the 2006 Constitution for which implementation had been long delayed.[73] The DRC's 2002 Mining Code has been under revision since 2012, but in March 2016 the government announced that the revision had been suspended until the recovery of metal prices.[74] Few measures are included in the 2002 mining legislation for consulting communities. Article 281 of the Code states that:

> [A]ny occupation of land that deprives the rights holders of enjoyment of the ground, any modification that renders the land unsuitable for cultivation brings about, for the titleholder or lessee of the mining and/or quarry rights, at the request of the land's rights holders and at their convenience, the obligation to pay fair compensation that corresponds either to the home, or to the value of the land at the time of its occupation, to which is added half of its value.[75]

The Code also does not contain measures for punishing companies that do not respect the provision on consulting communities.[76] In Katanga, companies like Chemaf and Ruashi Mining have failed to respect even these few measures for consultation and compensation. The gaps in the legal framework (such as the failure, described above, to define rights to the range of resources utilised by local communities), are problematic. Arguably a bigger problem is companies' failure to respect the provisions that are there, and the central government's inability or unwillingness to ensure they do so.[77]

As in the DRC, a key factor underpinning the persistence of violence in and around LSM sites in Zambia is the lack of recognition of certain types of claims, such as those based on customary land tenure or historical occupation by artisanal miners, and the incomplete enforcement of associated rights. Overlapping claims can arise when land dwellers hold land in the form of customary tenure, which is not considered legal title by the Zambian government. In 2010, for example, it was reported that individuals with guns, prospecting for manganese in Luapula Province (northern Zambia), had forced small-scale farmers off their land. A local NGO reported that they were receiving about 20 complaints per week, and that since 2009 over 500 people had been evicted from homes in the provincial capital. In Zambia the state possesses the right to invalidate claims to land granted by customary authorities – particularly given that the mining legislation prioritises mining over other uses.[78]

In the presence of resistance to dispossession or of disputes over compensation, violence would not be expected where 'there is government specification and enforcement of property rights. Here the government allocates and enforces rights, reducing the potential for conflict because we assume that governments have a comparative advantage in violence.'[79] An outcome in which governments were willing and able to fully enforce the PR granted to companies would be potentially highly exploitative, but likely also peaceful (in the sense of an absence of violence). Equally, if a government were willing to uphold artisanal miners' and communities' formal or informal claims vis-à-vis mining companies, these corporations would have to bargain with communities or miners to ensure a mutually acceptable outcome.

The persistence of violence in a context of limited state enforcement highlights the persistence of overlapping, unresolved claims to resources, for costly violent competition for resources demonstrates that peaceful agreement through bargaining and compensation could not be reached. Examining the dynamics that underpin the presence or absence of violence over mineral resources may shed light on the distributional implications of a given arrangement, for those left worse off may continue to press their claims, potentially leading to violence. Within a given community, certain members may be better off as a result of corporate investment, while others are unequivocally worse off. Hönke and Thomas argued that large Western mining companies in Katanga provided security around mining sites as a club good, making parts of the population more insecure.[80] The failure to compensate those losing access to resources and ensure a distributional outcome viewed as legitimate has the potential to result in lasting conflict and possibly violence.

Multinational corporations bring to bear a preponderance of resources compared to artisanal miners and communities. The case of the Kamoto Copper Company (KCC) and MUMI Mines (two subsidiaries of international company Glencore Plc) is an example of heavier reliance on coercive power (first-party enforcement) than second-party, coalitional mechanisms. Different actors were called upon to provide security: internal staff, private security companies, and the Congolese police at KCC and, at MUMI, the army. In June 2014 it was reported that on different occasions the mine police had shot at artisanal miners at KCC using live ammunition, leading to deaths and injury. One miner died in February 2014 as the result of a beating, having been captured by one of KCC's security patrols. Glencore has argued it does not control the Congolese mining police, despite the fact that they are paid by KCC. The authorities in Kolwezi have failed to investigate miners' deaths.[81]

In the DRC, limited government enforcement of companies' legal rights means that companies rely on a range of actors, including private security, to enforce their claims. CSR – often theorised on its own[82] – can be viewed as a strategy to exclude through an 'informal norm-based order or a coalition of interests'.[83] At one mine in Katanga, Hönke was told by the manager that

> [t]he idea is to make the poor village communities partners of the company. Once they come to see their own prosperity linked to that of the company, they should put social sanctions on theft from the mine and denounce intruders, even from their own villages.[84]

These norm-based arrangements are predicated on the assumption that companies and communities can receive shared benefits.

Coalitions of interests[85] or 'enforcement coalitions' will consist of the minimum of actors required to secure a company's claim. There is no institutional recipe; these coalitions could include local leaders, minority groups, community members, women, and more. The company seeks a coalition sufficiently strong to ensure the claim is respected. The creation of such coalitions can be based on norms, but also 'purchased' through payments or benefits to local officials or employment for specific individuals or groups – or obtained through coercion. Officials and agencies of the state – such as local officials or security forces – are often called upon to help enforce companies' claims.[86] In the DRC, state actors' involvement – often based on private interests[87] – can be viewed as second-party measures, rather than the legitimate state action of classical property rights theory. Companies draw on a range of context-specific strategies to enforce their claims, which can include CSR, payments or benefits to officials, pressure tactics, the hiring of certain individuals or groups, the selective distribution of benefits, and more. If an enforcement coalition is not strong or sustainable, it may break down – in which case the claim will be easily challenged,[88] potentially leading to more contestation. On the other hand, Hönke found in Katanga that all the bigger international mining firms drew on coercive enforcement.[89] Companies may be more likely to employ hard strategies if they expect resistance; if communities resist this enforcement, hard strategies may be more likely to result in violence.[90] While prospecting is taking place, the holder of mining rights and the owner of the surface may coexist with few problems. When mining activities are launched, the likelihood of disputes increases.[91]

Despite companies' range of tools and strategies, research in the DRC has demonstrated that MNCs' exclusionary capacity remains incomplete, resulting in repeated confrontations between artisanal miners and security forces.[92] At different mining sites in the DRC miners and communities continue to exercise informal rights. Companies are not the only ones to draw upon first-, second- and third-party mechanisms to enforce their rights to property; so do artisanal miners, highlighting a capacity for resistance that is often overlooked. Reports from Glencore's mining concession demonstrate that mine police regularly receive bribes and allow artisanal miners to enter the site.[93] For a fee, mine police officers and security guards let ASM miners know when it is safe to enter the site and when there is likely to be a security inspection. Yet these types of informal arrangements are fragile; if a security inspection takes place unexpectedly or police officers are dissatisfied with the fee paid by miners, violence may result.[94] Conflict may result with different actors excluded from enforcement coalitions; outright violence is by no means the only form of conflict in mining zones. Government repression of civil society actors working on mining issues, for example, has been documented in the DRC.[95]

In the DRC, the 2002 Mining Code was created to favour the development of the private sector, particularly LSM. The Congolese Code and later Mining Regulations distinguish between LSM, ASM and small-scale mining. Yet Congolese law has little to say about ASM. The Ministry of Mines can establish artisanal exploitation zones (AEZs) if, for technological and economic reasons, industrial extraction is viewed as inappropriate. Companies cannot be granted research permits within the boundaries of an AEZ, unless a cooperative of artisanal miners requests it. However, the law allows for the AEZ to be shut down if there is a change in the factors that led to its creation, or a new deposit is discovered that requires large-scale extraction. Therefore, in reality, LSM interests can trump ASM. Artisanal miners must purchase an official authorisation card before they can mine in an AEZ; the card has to be renewed annually, which reduces tenure security.[96] There is also a gap between the law and its implementation. As of 2012 few AEZs had been created; those that had been created had proven unsuitable for ASM, and were not even operational.[97]

An argument raised with respect to continued poor standards of living in communities surrounding mine sites is that the reputational or other costs to companies are not high enough for them to invest resources.[98] Yet we could therefore expect all mining sites to be equally conflictual or violent, which is not the case. This fact merits in-depth, comparative study of the enforcement coalitions at specific sites, and how these coalitions are configured and reconfigured throughout the life of a mine.[99] Observing the mine cycle is important in the context of dynamic investment in the DRC and Zambia, in which asset sales, joint ventures, and contract revisions are commonplace.

Conclusion

Classical property rights theory has limitations when it comes to shedding light on LSM in Zambia and the DRC. Many of the dynamics around mining investment are not well captured by the predictions of early PR theorists, whose work was developed to explain the emergence of private property in Europe and North America. To argue that the reality in Zambia and the DRC is different is not to buy into modernisation approaches according to which the Global South is (or should be) evolving towards European or North American 'modernity'. Rather, it is to shine light on the fact that PR transformation is messy and unlikely to be as simple as a government setting aside a large concession for a mining company. The persistence of contestation and conflict around a number of mining sites is a sign that this messiness is likely to persist, companies' efforts to fine-tune their CSR notwithstanding.

Classical property rights theory – and, for a long time, policy advice on agricultural land development in the Global South – touted the efficiency benefits of private property. Clearly defined, secure rights, it was believed, would encourage subsistence farmers on communally held land to increase investment. Greater investment would bring increased productivity and enhanced food security nationally. Yet due to PR dynamics in contexts of legal pluralism, contestation around LSM sites in the DRC and Zambia may persist. This is particularly the case given the continued importance of customary authority for the allocation of land; this authority has in some places, like Zambia's 'new Copperbelt', been *reinforced* by LSM investment.[100]

Classical property rights theory is predicated on a belief in the greater efficiency of clearly defined, well-enforced PR. National policy-makers make similar assumptions regarding the larger overall contribution to the economy from granting PR to LSM enterprises, particularly compared to the seeming absence of clearly defined rights and responsibilities at ASM sites. This preference underpins legislative provisions like dominant interest. International financial institutions like the World Bank and the IMF were instrumental in pushing for LSM-friendly mining codes in countries like the DRC.[101] Yet a large and growing body of scholarly research highlights the organised nature of artisanal mining, even when communal rather than private PR are the norm.[102] Such assumptions should therefore be questioned, given that in countries like the DRC, LSM revenues are unlikely to be reinvested.

International development donor initiatives have emphasised the need to formalise ASM, with the expectation of national development through increased taxation. USAID's Property

Rights and Artisanal Diamond Development (PRADD) project in the Central African Republic and Liberia, and PRADD II in Côte d'Ivoire and Guinea, worked towards the identification and formalisation of property rights in artisanal diamond-mining areas through measures like the reduction of licensing, royalty, tax, and fee costs, with some success.[103] The question raised throughout this paper, however, remains: whether technical measures can fundamentally resolve the deeply political nature of overlapping claims to resources. The modified property rights framework presented here, which takes into account overlapping claims to resources and how they are enforced, can help to make sense of peace and conflict in the DRC and Zambia, and potentially other resource-rich African countries as well.

Disclosure statement

No potential conflict of interest was reported by the author.

Notes

1. Libecap, "Property Rights in Economic History," 227.
2. Meinzen-Dick and Mwangi, "Cutting the Web of Interests," 36.
3. Ibid., 36.
4. Risse, *Governance Without a State?*
5. Risse and Lehmkuhl, "Governance in Areas of Limited Statehood," 4.
6. Fitzpatrick, "Evolution and Chaos," 998; and Boserup, *The Conditions of Agricultural Growth*.
7. Mwangi, "Subdividing the Commons," 817.
8. Acemoglu and Robinson, *Why Nations Fail*; and Libecap, "Property Rights in Economic History."
9. Acemoglu and Robinson, *Why Nations Fail*.
10. Meinzen-Dick and Mwangi, "Cutting the Web of Interests," 37.
11. Fitzpatrick, "Evolution and Chaos," 1000.
12. Ibid., 1000.
13. Alston et al., "De Facto and De Jure Property Rights," 5.
14. Ibid., 47.
15. Ibid., 2.
16. Ibid., 10.
17. See note 4 above.
18. See note 9 above.
19. Mamdani, Citizen and Subject.
20. PREMICONGO, *Les investissements miniers chinois*.
21. Mamdani, "Citizen and Subject."
22. Fitzpatrick, "Evolution and Chaos," 999.

23. Bruce and Mearns, *Natural Resource Management*.
24. Fitzpatrick, "Evolution and Chaos," 1047.
25. Mwangi, "Subdividing the Commons," 815.
26. Ibid., 815.
27. Ibid., 817.
28. Ibid., 817–8.
29. Ibid., 818.
30. North, *Institutions, Institutional Change*.
31. E.g. The World Bank, "Small-scale Mining," http://web.worldbank.org/WBSITE/EXTERNAL/TOPICS/EXTOGMC/0,,contentMDK:20246087~menuPK:509392~pagePK:148956~piPK:216618~theSitePK:336930~isCURL:Y,00.html.
32. Geenen, "A Dangerous Bet," 323.
33. Ibid., 323–4.
34. Ibid., 324.
35. See note 20 above.
36. Agarwal, "Gender and Land Rights Revisited."
37. ActionAid, "Women in Zambia."
38. Geenen and Hönke, "Land Grabbing by Mining Companies."
39. Bolay, "When Miners Become Foreigners."
40. Fleury and Davies, "Sustainable Supply Chains."
41. Mwangi, "Subdividing the Commons," 816.
42. Geenen, "A Dangerous Bet," 322.
43. The Carter Center, *Les investissements miniers*.
44. Ndulo, *Legal and Regulatory Frameworks*.
45. Ibid., 7.
46. Ndulo, "Mining Legislation and Mineral Development in Zambia," 13–14.
47. Ndulo, *Legal and Regulatory Frameworks*, 7.
48. Ibid., 7.
49. Ibid., 8.
50. Ndulo, "Mining Legislation and Mineral Development in Zambia," 19.
51. Ndulo, *Legal and Regulatory Frameworks*, 8.
52. Ibid., 9–10.
53. Ibid., 12.
54. Ibid., 13.
55. Ibid., 20.
56. Ibid., 15, emphasis added.
57. Ibid., 15.
58. Ibid., 14.
59. The Carter Center, *Les Investissements Miniers*, 22.
60. Ibid., 22, my translation.
61. Libecap, "Property Rights in Economic History," 235.
62. E.g. USAID, Property Rights and Artisanal Mining – Clarifying and Strengthening Rights: Options for Policymakers, http://www.usaidlandtenure.net/sites/default/files/USAID_Land_Tenure_Artisanal_Mining_Issue_Brief_0.pdf.
63. E.g. African Development Bank Group, "Delivering on the Promise: Leveraging Natural Resources to Accelerate Human Development in Africa," http://www.afdb.org/en/topics-and-sectors/initiatives-partnerships/african-natural-resources-center-anrc/delivering-on-the-promise-leveraging-natural-resources-to-accelerate-human-development-in-africa/.
64. Libecap, "Property Rights in Economic History," 227–8.
65. Alston et al., "De Facto and De Jure Property Rights," 10.
66. See note 64 above.
67. Amnesty International, *Profits and Loss*, 22–3.
68. Ibid., 13.
69. Ibid., 14.

70. Ibid., 15.
71. Ibid., 15–16.
72. Ibid., 21–2.
73. *Radio Okapi*, August 18, 2015.
74. Reuters, *Congo Mining Code Revision*.
75. Ibid., 22, my translation.
76. The Carter Center, *Les Investissements Miniers*, 20, 22.
77. Ibid., 20.
78. IRIN, *Zambia: Land and Mineral Rights in Conflict*.
79. Alston et al., "De Facto and De Jure Property Rights," 6.
80. Hönke and Thomas, *Governance for Whom?* 16.
81. Peyer et al., "PR or Progress?" 10–11.
82. E.g. *Resources Policy* 37, no. 2 (2012).
83. See note 11 above.
84. Hönke, *Transnational Pockets of Territoriality*, 19.
85. See note 11 above.
86. Hönke, *Transnational Pockets of Territoriality*, 16.
87. Hönke, "New Political Topographies," 9.
88. Ibid.
89. See note 86 above.
90. Steinberg, "Strategic Sovereignty."
91. Ndulo, *Legal and Regulatory Frameworks*, 13.
92. Hönke, "New Political Topographies," 10.
93. Peyer et al., "PR or Progress?" 61.
94. Peyer et al., "PR or Progress?" 51.
95. PREMICONGO, *Les investissements miniers chinois*, 26; and Kairos, "Congolese Human Rights Defender."
96. Geenen, "A Dangerous Bet," 325.
97. Ibid., 326.
98. Ibid., 94.
99. Fleury and Davies, "Sustainable Supply Chains."
100. Negi, "The Mining Boom," 210.
101. Geenen, "A Dangerous Bet," 325; and Campbell, "Revisiting the Reform Process."
102. Geenen, "A Dangerous Bet."
103. Hinton and Levin, "Property Rights and Artisanal Diamond Development," v.

Bibliography

Acemoglu, Daron, and J. A. Robinson. *Why Nations Fail: The Origins of Power, Prosperity, and Poverty*. New York: Crown Business, 2012.

ActionAid. "Women in Zambia Bear the Brunt of the Harmful Impact of Mining." ActionAid Zambia, 2015. http://www.actionaid.org/zambia/news/women-zambia-bear-brunt-harmful-impact-mining.

Alston, L. J., Edwyna Harris, and Bernardo Mueller. *De Facto and De Jure Property Rights: Land Settlement and Land Conflict on the Australian, Brazilian and U.S. Frontiers*. Cambridge, MA: National Bureau of Economic Research, 2009.

Amnesty International. *Profits and Loss: Mining and Human Rights in Katanga, Democratic Republic of Congo*. London: Amnesty International, 2013. https://www.amnesty.org/en/documents/afr62/001/2013/en/.

Bolay, Matthieu. "When Miners Become "Foreigners": Competing Categorizations within Gold Mining Spaces in Guinea." *Resources Policy* 40 (2014): 117–127. doi:http://dx.doi.org/10.1016/j.resourpol.2014.02.001.

Boserup, Ester. *The Conditions of Agricultural Growth: The Economics of Agrarian Change under Population Pressure*. London: George Allen & Unwin, 1965. http://www.biw.kuleuven.be/aee/clo/idessa_files/boserup1965.pdf.

Bruce, J. W., and Robin Mearns. *Natural Resource Management and Land Policy in Developing Countries: Lessons Learned and New Challenges for the World Bank*. International Institute for Environment and Development, 2002. http://pubs.iied.org/9222IIED.html.

Campbell, Bonnie. "Revisiting the Reform Process of African Mining Regimes." *Canadian Journal of Development Studies* 30 (2010): 197–217. doi:http://dx.doi.org/10.1080/02255189.2010.9669288.

Fitzpatrick, Daniel. "Evolution and Chaos in Property Rights Systems: The Third World Tragedy of Contested Access." *The Yale Law Journal* 115 (2006): 996–1048. http://www.yalelawjournal.org/essay/evolution-and-chaos-in-property-rights-systemsthe-third-world-tragedy-of-contested-access.

Fleury, A.-M., and Benjamin Davies. "Sustainable Supply Chains – Minerals and Sustainable Development, Going Beyond the Mine." *Resources Policy* 37, no. 2 (2012): 175–178. doi: http://dx.doi.org/10.1016/j.resourpol.2012.01.003.

Geenen, Sara. "A Dangerous Bet: The Challenges of Formalizing Artisanal Mining in the Democratic Republic of Congo." *Resources Policy* 37, no. 3 (2012): 322–330. doi:http://dx.doi.org/10.1016/j.resourpol.2012.02.004.

Geenen, Sara, and Jana Hönke. "Land Grabbing by Mining Companies: Local Contentions and State Reconfiguration in South Kivu (DRC)." In *Losing your Land: Dispossession in the Great Lakes*, edited by A. Ansoms and T. Hilhorst, 58–81. Oxford: James Currey, 2014.

Hinton, Jennifer, and Estelle Levin. *Property Rights and Artisanal Diamond Development (PRADD) Project Comparative Study: Legal and Fiscal Regimes for Artisanal Diamond Mining*. USAID, 2010. http://www.usaidlandtenure.net/praddcar/comparative-study.

Hönke, Jana. *Transnational Pockets of Territoriality. Governing the Security of Extraction in Katanga (DRC)*. 2009. http://www.uni-leipzig.de/~ral/gchuman/fileadmin/media/publikationen/Working_Paper_Series/RAL_WP_2_Hoenke_web.pdf.

Hönke, Jana. "New Political Topographies: Mining Companies and Indirect Discharge in Southern Katanga (DRC)." *Politique Africaine* 120, no. 4 (2010): 105–127. doi:http://dx.doi.org/10.3917/polaf.120.0105.

Hönke, Jana, and Esther Thomas. *Governance for Whom? Capturing the Inclusiveness and Unintended Effects of Governance*. Berlin: DFG Collaborative Research Center, 2012. http://www.research.ed.ac.uk/portal/en/publications/governance-for-whom(a5c395ae-fdd2-4d45-bf6b-19f11b5dcbed).html.

IRIN. *Zambia: Land and Mineral Rights in Conflict*. 2010. http://www.irinnews.org/report/90338/zambia-land-and-mineral-rights-in-conflict.

Kairos. "Congolese Human Rights Defender Imprisoned for Questioning Mining Practices." Kairos. Accessed March 23, 2016. http://www.greysisters.ca/engagingwithourworld/currentissues/documents/GoldenMisabiko.pdf

Libecap, G. D. "Property Rights in Economic History: Implications for Research." *Explorations in Economic History* 23 (1986): 227–252. http://www.colorado.edu/ibs/es/alston/econ4524/readings/Libecap,%20Property%20Rights%20in%20Economic%20History.pdf.

Meinzen-Dick, Ruth, and Esther Mwangi. "Cutting the Web of Interests: Pitfalls of Formalizing Property Rights." *Land Use Policy* 26, no. 1 (2009): 36–43. doi:http://dx.doi.org/10.1016/j.landusepol.2007.06.003.

Mwangi, Esther. "Subdividing the Commons: Distributional Conflict in the Transition from Collective to Individual Property Rights in Kenya's Maasailand." *World Development* 35, no. 5 (2007): 815–834. doi: http://dx.doi.org/10.1016/j.worlddev.2006.09.012.

Ndulo, Muna. "Mining Legislation and Mineral Development in Zambia." *Cornell International Law Journal* 19, no. 1 (1986): 1–34. http://scholarship.law.cornell.edu/cilj/vol19/iss1/1/.

Ndulo, Muna. *African Development Bank High Level Policy Seminar on Optimizing the Benefits of Coal & Gas in Mozambique: Legal and Regulatory Frameworks for Resource Exploration and Extraction-Global Experience*. 2013. http://www.afdb.org/fileadmin/uploads/afdb/Documents/Generic-Documents/Presentation%20-%20Legal%20and%20Regulatory%20Frameworks%20for%20Resource%20Exploration.pdf.

Negi, R. "The Mining Boom, Capital, and Chiefs in the 'New Copperbelt'." In *Zambia, Mining, and Neoliberalism: Boom and Bust on the Globalized Copperbelt*, edited by A. Fraser and M. Larmer, 209–236. Palgrave Macmillan, 2010.

North, D. C. *Institutions, Institutional Change and Economic Performance*. Cambridge, UK: Cambridge University Press, 1990.

Peyer, Chantal, Patricia Feeney, and François Mercier. *PR or Progress? Glencore's Corporate Responsibility in the Democratic Republic of the Congo*. Bread for All, Fastenopfer, and RAID, 2014.

PREMICONGO. *Les Investissements Miniers Chinois au Katanga et la Détresse des Communautés Locales: Cas de la Minière de Kalumbwe Myunga (MKM) et de HUACHIN* [Chinese mining investments in Katanga and local communities' distress: The case of Kalumbwe Myunga Mining (MKM) and HUACHIN]. 2015. http://www.congomines.org/system/attachments/assets/000/000/925/original/Les_investissements_chinois_au_Katanga.pdf?1447927146.

Reuters. *UPDATE 1-Congo Mining Code Revision Suspended Until Metal Prices Recover –Minister*. 2016. http://af.reuters.com/article/drcNews/idAFL5N16I5EV.

Risse, Thomas, ed. *Governance Without a State? Policies and Politics in Areas of Limited Statehood*. New York: Columbia University Press, 2011.

Risse, Thomas, and Ursula Lehmkuhl. *Governance in Areas of Limited Statehood – New Modes of Governance?* SFB-Governance Working Paper Series, No. 1. Berlin: Research Center (SFB), 2006. http://www.sfb-governance.de/publikationen/working_papers/wp1_en/SFB-Governance-Working-Paper-1_en.pdf.

Steinberg, Jessica. "Strategic Sovereignty: A Model of Non-state Goods Provision and Resistance in Regions of Natural Resource Extraction." *Journal of Conflict Resolution* (2015). doi:http://dx.doi.org/10.1177/0022002714564429.

The Carter Center. *Les Investissements Miniers en République Démocratique du Congo: Développement ou Appauvrissement des Communautés Locales?* [Mining investments in the Democratic Republic of Congo: Development or Impoverishment of Local Communities?] 2012. https://www.cartercenter.org/resources/pdfs/ news/peace_publications/human-rights/carter-center-mining-governance-HRIA-Oct2012.pdf.

USAID. *Program Brief: Property Rights and Artisanal Diamond Development (PRADD)*. 2012. http://www.usaidlandtenure.net/content/program-brief-property-rights-and-artisanal-diamond-development-pradd.

'Custom' and fractured 'community': mining, property disputes and law on the platinum belt, South Africa

Sonwabile Mnwana

ABSTRACT
Drawing on a recent study carried out in the Bakgatla-ba-Kgafela tribal authority area in South Africa's North West province, this article demonstrates that rural-based platinum mining generates intense struggles over 'communal' property – land and mining revenues. Competing versions of custom and contrasting group identities dominate these struggles. Since rights to communal property are secured through group membership and defined through custom, custom itself has become a space for contestation. It is the inflexibility in the way the state and the courts conceive custom – as a rigid set of rules – which perpetuates inequality and disputes at village level.

Introduction

In this article, I draw on a recent study carried out in the Bakgatla-ba-Kgafela tribal authority area in South Africa's North West province, to illustrate how the structure of power seems to shape the character of resistance on the platinum-rich communities. This article demonstrates how platinum mining generates intense rural struggles over communal property – land and mining revenues. Some scholars observe that the escalating conflict on the platinum belt epitomises a resurgence of ethnic identities in post-apartheid South Africa.[1] However, I argue that competing versions of custom and contrasting group identities that dominate these struggles reveal not just ethnic revitalisation, but serious divisions that challenge the very meaning and existence of 'tribal (traditional) communities'. Since rights to communal property are secured through group ('tribal') membership and defined through custom, custom itself has become a space for contestation. Therefore, the colonial crystallisation of custom made it difficult to 'bend' when people's social conditions change, but easy to 'break' when disputes over property arise. It is this inflexibility in the way the state and the courts conceive custom – as a rigid set of rules – which perpetuates inequality and disputes at village level.

As it surfaces later in the empirical section, this article illustrates this argument through summaries of court disputes over land mining revenues between the Bakgatla chiefs and ordinary villagers. Since the mid-1990s when mining expansion intensified in the Bakgatla area, disputes over the mineral-rich land (farms), mining revenues and political power have also escalated. On several occasions these disputes end up in the courts of law. The courts' interpretation of custom in the various judgements significantly favours the chiefs' powers and control over communal property. I discuss five of these court cases in this contribution. To bring a historical perspective, I discuss the 1956 legal case between *Kgosi* Tidimane and a village activist, Jacob Pilane who filed a court petition against the chief for his failure to account for the money collected for purchasing certain farms. I then discuss some of the recent court battles between Kgosi Nyalala Pilane and village activists in Lesetlheng, Motlhabe and Sefikile villages in the Bakgatla area. In all these cases the courts' interpretation of customary law remains similar, particularly when it comes to the power of chiefs over communal property.

Historical crystallisation and distortion of custom

Land is the basic means by which Africans in rural areas can access property and the natural resources they need for survival. As rural land becomes increasingly scarce and its value increases, competition and conflict over it intensify in different parts of sub-Saharan Africa and new forms of exclusion, inequalities and differentiation have been reported.[2] Struggles over land however are, in essence, broader wrangles over property – particularly social relations over it and power and legitimacy of local political authorities regulating it.[3] Rights to property – primarily land – are often defined as customary or communal. Power and legitimacy of property regulating governing authorities – mainly 'traditional authorities' – in Africa are mobilised and defined in terms of custom or the so-called customary law.[4]

As elsewhere in most postcolonial sub-Saharan Africa the land tenure rights of indigenous[5] communities in South Africa's former 'homelands'[6] are extremely precarious and ambiguous. The legal status and tenure security of the land occupied and used by Africans, especially individual families in rural areas, remains unclear. At the root of this paradox lies the history of dispossession of indigenous communities and the imposition of significantly distorted versions of 'custom' (customary law) during colonial apartheid periods. This issue is further complicated by the state and other actors in South Africa who generally define African property rights as 'communal'.[7] As such, tenure rights to land and other natural resources are generally conceived as communal – shared by groups who are defined as political units – 'tribes'.

But the content of African custom was significantly reshaped or distorted by colonial and apartheid experience in South Africa and elsewhere in Africa. The colonial process of 'formalisation' or crystallisation of custom enhanced the power of chiefs. For Mamdani, customary law became both 'all embracing' and divisive. It 'embraced' under the power of chiefs 'previously autonomous social domains [among others] the household, age sets, and gender'. Yet, the purpose of customary law, argues Mamdani, 'was not about guaranteeing rights, it was about enforcing custom. It was not about limiting the power [of chiefs], but about enabling it'.[8] As such, the rigid, crystallised, ubiquitous version of custom which conceptualised African custom as law – legally enforceable rules that governed all aspects of African life – has no pre-colonial precedent. Chanock, for instance, maintains that

> [in] Africa, as elsewhere, much of life is lived outside of the law and involves values and patterns of behaviour which are different from these enshrined in the state's legal system. In these patterns may be found a repository of customary values which provide better building blocks than those which were legitimated by the colonial order.[9]

The wholesale denial of private property rights and enforcement of chiefly mediated 'communal' rights and tribal identities to Africans was an integral part of the system of indirect rule employed by the colonial state.[10] This system was primarily influenced and justified by the social Darwinist school of thought which greatly shaped the administrative and policy thinking in the twentieth-century British colonial rule.[11] Influenced by this strand of thinking, argues Delius, colonial officials perceived land rights within African population to be 'communal' in nature since African peoples were seen 'as being at a lower level of social evolution'. Thus,

> private property was the mark of civilisation while less evolved societies were believed to have weak communal rights.[12]

Social Darwinism was applied vigorously in the colonial state's late nineteenth- and twentieth-century land policies that were directed at addressing the African land tenure question in South Africa. Such a policy practice and ideological premise also promoted the faulty notion that all Africans were chiefly subjects and thus members of tribes. In South Africa, the colonially derived versions of customary rights and governance mechanisms, albeit legally ambivalent and administratively laborious, were to be kept intact and sustained through colonial declarations and the two major 'Native' Land Acts of 1913 and 1936.[13]

Colonial laws in South Africa rendered it impossible for the rural poor in the former homelands to have communal tenure without the control of tribal authorities. Such a belief is not only ahistorical, but also groundless. In pre-colonial times many African groups and individual families held land and other property rights outside the boundaries of traditional institutions and custom.[14] Moreover, literature on the pre-colonial history of Africa offers some evidence that key factors such as availability of land and other mechanisms which ensured downward accountability of pre-colonial chiefs made it easy for people to leave the chiefdoms of despotic or weak chiefs and join other 'tribes'. Thus, as Delius puts it:

> The availability of land made it relatively easy for groups to move between chiefdoms. The evidence suggests that there was considerable movement between chiefdoms in 18th and 19th century South Africa, and that this process placed important checks on chiefly abuse of power. This mobility also contributed to cultural heterogeneity, which undermines the depiction of these communities as tribes ... In fact, the populations of many chiefdoms were diverse and their boundaries were porous.[15]

The colonial state significantly empowered local chiefs. The latter were assumed to be custodians of tribal properties. Subsequently, chiefs had enormous leverage to define and impose customary rights on rural residents (their 'subjects'). Therefore, custom became an instrument of colonial administration, control and coercion over African masses.[16] By and large, rural power was organised at the level of a local institution – a 'tribal authority'. Colonial state's power over Africans, decentralised to the level of chiefs, had to be legitimised and defended through the distorted versions of tradition and custom. Chiefs' power became solidified over 'communal' property to crush rural resistance against the state.[17] Even on countless occasions where Africans purchased landed property privately as distinct groups to avoid state-imposed 'tribal' identities and custom, colonial policies enforced a requirement that they should constitute themselves as 'tribes' and register their properties under the names of a 'recognised' chief. Such a property was then transferred to a state 'trustee' – a colonial official – who held the title deed in trust for the chief and his tribe.[18]

Central to the colonial mode of crystallising and enforcing customary law were the courts. For Mamdani, it was the courts – 'the torchbearers of that civilization'[19] – who enforced customary law. Law was central to colonialism's chief claim of bringing civilisation to Africa. By civilisation the colonial state meant 'the rule of law'.[20] However, continues Mamdani, the colonial mandate could not align with 'civilizing' courts of law. As such, the mandate shifted from 'the civilizing mission to law and order administration'.[21]

The post-apartheid Constitution[22] in South Africa recognises customary law and mandates the courts to:

> [A]pply customary law when that law is applicable, subject to the Constitution and any legislation that specifically deals with customary law.[23]

However, this mandate seems difficult to realise in the light of post-1994 legislation that reinforces the apartheid-style power and authority of chiefs. Claassens cautions:

> [T]o determine the content of customary law by standards of 'formal' law is to apply a distorted paradigm.[24]

Generally, the courts in South Africa have increasingly categorised custom into two main categories: the 'official' and the 'living' law. 'Official' customary law is a product of the state and legal experts,[25] while 'living' law refers to the law actually observed by the people who created it'.[26] Official customary law is a product of colonial formalisation of indigenous peoples' law, which imposed rigid, Western, rule-oriented conceptions of law and order. Living law, on the other hand, evolves organically out of ever-changing African sociocultural 'processes' of dispute resolution.[27] Therefore, it is through the process of colonial formalisation that African 'living law' became distorted.

Platinum mining, communities and chiefly power

South Africa is the world's largest producer of platinum.[28] Increasingly, land which falls under the former 'homelands' of South Africa has become the target of the post-apartheid mining focus. This emanates from the country's geological advantage – the vast platinum-rich rock formation called the Igneous Bushveld Complex (the Bushveld Complex). Most of the area covered by the Bushveld Complex – tagged 'the platinum belt' – spreads beneath rural communal land under the political jurisdiction of traditional (formerly known as 'tribal') authorities[29] mainly in the North West and Limpopo provinces. The two provinces fell under the 'independent homelands' of Bophuthatswana and Lebowa, respectively, during apartheid. As such, they bear the marks of the oppressive apartheid order: extreme poverty, massive unemployment, poor education and a lack of basic public services. The leading global platinum producers such as Anglo American Platinum Limited (Amplats), Impala Platinum Holdings Limited (Implats) and Lonmin Plc (Lonmin) compete for space with communities in these overcrowded areas. The rural-based South African platinum industry has shifted the geographical focus of post-apartheid mining – its predecessor, the gold industry, largely took place in urban industrial centres.

The expansion of the mining industry in communal areas coincides with post-apartheid attempts to redefine residents in these areas, through law, as subjects of 'traditional communities' (or 'tribes') under chiefs. Legislation that has been enacted since the early 2000s has not only legitimised the mediation of mine-community relationships by traditional leaders, but has also significantly enhanced the powers of chiefs in South Africa. Although the post-1994 African National Congress government at first vacillated about defining and codifying the powers and status of chiefs, it eventually passed legislation that significantly increased the powers of chiefs in rural local governance. The Traditional Leadership and Governance Framework Act of 2003 (Act 41 of 2003, or the TLGFA)[30] is the main piece of legislation in this regard.

The TLGFA re-enacts traditional (tribal) authorities to preside over precisely the same geographic areas that were defined by the apartheid government.[31] Among other things, the Act enables chiefs and their traditional councils to be granted powers over the administration and control of communal land and natural resources, economic development, health and welfare and to administer justice.[32] As such, not only does this Act impose the former colonial tribal authority's demarcations on rural citizens, it also promotes a controversial governance role for chiefs. Other controversial laws that, so far, have been successfully resisted by rural citizens include the Communal Land Rights Act[33] of 2004 (Act 11 of 2004)[34] and the Traditional Courts[35] Bill.[36]

Post-apartheid laws regulating mineral rights, particularly the Minerals and Petroleum Resources Development Act of 2002 (Act 28 of 2002) and its accompanying regulations, also drives the inclusion of traditional communities in South Africa's platinum industry. In seeking to redress past injustices by transforming relationships between the mining companies and local communities, this legislation has adopted a range of measures, including continued royalty payments, black economic empowerment mine-community partnerships and social labour plans as requirements for mining companies. The state has encouraged communities[37] who previously received royalty compensations for loss of land due to mining, to convert their royalties into equity shares. Consequently, with the state's support, chiefs, as assumed custodians of communal resources, have become mediators of mineral-led development and mining deals. The Bakgatla ba Kgafela community is one of the communities on the platinum belt that has significant control over mining revenues as a result of mining in the area. Other communities with stakes in platinum mining in North West include the Bafokeng, Bakubung ba Rutheo, Bapo ba Mogale and baKwena ba Mogopa.

This means that traditional communities' interactions and engagements with mining companies are mediated and controlled by local chiefs. As assumed custodians of rural land and other tribal properties, chiefs enter into mining contracts and receive royalties and dividends on behalf of rural residents who live in the mineral-rich traditional authority area. This traditional-elite mediated model of community participation in the mining industry[38] has received increased media attention,[39] particularly since the 2012 Marikana massacre.[40] Members of the South African Police Service shot at striking mine workers at the Lonmin Mine, North West province, and the massacre left 34 miners dead and many injured.

In the face of protracted labour unrest in the platinum sector and the recent sharp decline in commodity prices, the dominant view propagated by government, mining companies and the chiefs is that tribal-elite mediated community control of mineral revenues is crucial for congenial relations within the rural-based platinum sector. Chiefs see themselves as legitimate mediators and gatekeepers through whom mining capital can gain 'easy access' to cheap local labour and communal land. However, recent research has shown that this model has not yet led to tangible benefits for community members, rather, it has enhanced the power of the chiefs and led to a lack of transparency, unaccountability, heightened inequality, deepened poverty and local tensions.[41] Post-apartheid laws regulating and governing traditional leadership and mining reform have been criticised for promoting exclusion and corruption using 'distorted constructs of custom' to 'impose contested identities' and 'undermining [rural residents'] capacity to protect their land and … mineral rights.'[42]

A note on data collection

This study is part of ongoing research which explores the multiple impacts of platinum mining in South Africa's former homeland areas. A research team comprising myself (the researcher) and two research assistants collected data in three stages between July and November 2013 in three villages in the Bakgatla area: Motlhabe, Sefikile and Lesetlheng. The first stage was local-level fieldwork. Our methods included in-depth interviews, observations and analysis of documents (court case reports). The second stage involved several visits to the South African National Archives in Pretoria. The third stage was a detailed collection of oral histories. The main purpose was to focus on land and political histories and how they inform the present struggles in the Bakgatla area. During the different stages of data collection we participated in a range of activities, including *makgotla* (village meetings), youth gatherings, court cases and state-instituted commission hearings on Bakgatla chieftainship disputes. The study is still in progress and I continue to make sporadic follow-up research visits to the study area. The ethnographic material presented here is based on selected semi-structured key-informant interviews with village activists in the selected villages.[43] This selected ethnographic material is corroborated by reference to selected archival documents in the South African National Archives in Pretoria.

The Bakgatla-ba-Kgafela traditional authority area

A product of South Africa's turbulent history of colonialism and racial oppression during apartheid, the eccentric confederation of 32 villages that constitute the Bakgatla area spreads all over the north-eastern and north-western foothills of the magnificent Pilanesberg mountains. This largely Setswana-speaking traditional authority area, under the leadership of *Kgosi* (Chief) Nyalala Pilane, is one of the largest communal areas in the North West province. It covers a vast area of more than 40 farms in the Pilanesberg region, about 60 km north of the town of Rustenburg. With approximately 160,000 residents[44] the Bakgatla-ba-Kgafela area, which falls under the Moses Kotane Local Municipality (MKLM), is the epitome of a prominent tribal authority area with vast mineral resources. As is the case in many rural areas in South Africa, the relationship between the Bakgatla traditional authority and the MKLM has not been smooth. This can be attributed largely to the lack of legal or administrative clarity about the role of traditional leaders in 'developmental' local governance (Figure 1).[45]

Mining revenues and resistance

Like other tribal authorities that were incorporated into the Bophuthatswana 'homeland' in 1977, the Bakgatla chieftaincy had farms registered with the state in trust for the chief and his tribe. These farms spanned over some of the richest platinum group metals (PGM) resources on the western limb of the Bushveld Complex.

During apartheid, the Minister of Bantu Affairs, in his capacity as a 'tribal land trustee' mediated the contracts between mining companies and tribal authorities.[46] The powers of 'state trusteeship' regarding mineral and surface rights on tribal land were to be transferred to the Office of the President of Bophuthatswana[47] when the latter gained its 'independence'. It was during the Bophuthatswana regime that the Bakgatla chieftaincy began to receive mining royalties from Anglo American Platinum (Amplats's Union Mine) in 1982.[48] These

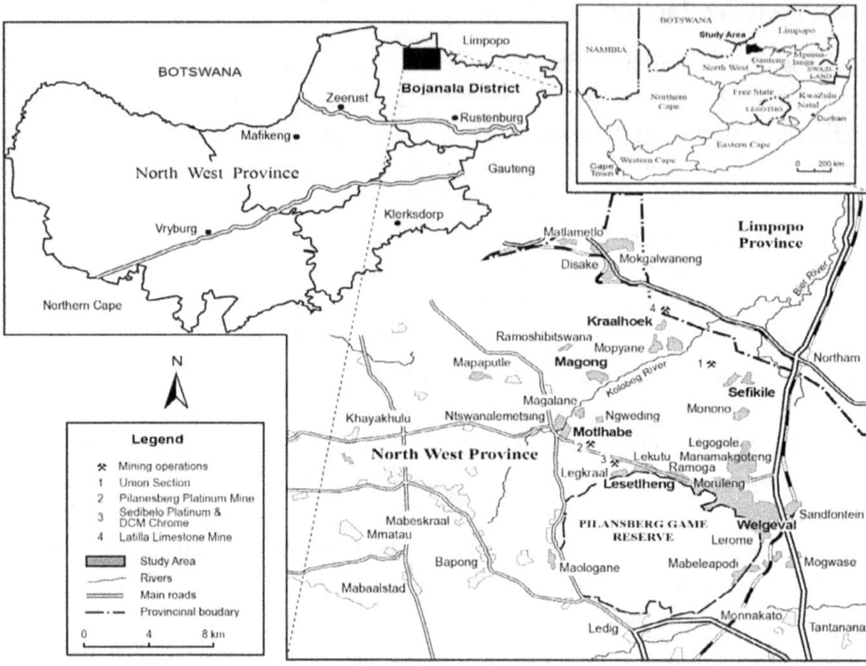

Figure 1. Map showing the Bakgatla-ba-Kgafela area.
Source: Society Work and Development Institute (SWOP), University of the Witwatersrand.

royalties were deposited into Bophuthatswana state accounts called D-accounts, held by the Office of the President. Due to the global upsurge in platinum demand in the 1980s and 1990s the royalties payable rose significantly. From time to time, the chiefs would request funds from these accounts for various community development projects. If granted, funds would be deposited into a local tribal trust account administered by a local state official. A tribal trust account for the Bakgatla community still exists at the District Office for Local Government and Traditional Affairs (LGTA) in Mogwase, about 5 km from Sun City. According to the manager at the LGTA office in Mogwase, in the 1980s and early 1990s mining royalties in Bakgatla Trust account amounted to between two and three million rand (roughly17357257.500 US Dollar). Under *Kgosi* Nyalala Pilane, the Bakgatla Traditional Administration has channelled various mining revenues to private business accounts that he opened in the community's name.[49]

After South Africa's first democratic elections in 1994, the D-accounts fell under the administration of the province's Department of Finance and LGTA. Since then, significant amounts of money have mysteriously disappeared from these accounts.[50] Since it took over the administration of these funds, the North West provincial government has been neither transparent nor accountable to the relevant communities.

The global boom in PGM prices, which began in the early 1990s, ushered the Bakgatla-ba-Kgafela area and other rural communities on the platinum belt to centre stage of South Africa's post-apartheid mining economy. Several mining operations have developed in Bakgatla-ba-Kgafela territory. On behalf of the residents in the area under his jurisdiction, *Kgosi* Nyalala Pilane has entered into numerous deals and concessions with the mining companies and other investors.[51] As a result of these deals, the Bakgatla-ba-Kgafela community has become a huge business empire worth approximately R15 billion (1,041,435,450 US Dollar).[52] This has heightened the chief's power and status.

The mounting resistance against *Kgosi* Nyalala Pilane is also due to his lack of transparency and accountability in corporate dealings, and various groups' allegations of corruption against him. The mining-related investments that the *kgosi* has created through contracts with the mining companies are legion. And they keep changing. This makes it difficult to keep up with them. *Kgosi* Nyalala Pilane is the director in most of these complex company structures that bear the Bakgatla name. Some village groups contest these mining contracts that are signed by the chief. They argue that their forefathers bought the mineral-rich farms as private properties and they should never have become tribal land. Most of the struggles in the Bakgatla area between the chief and various village groups have been fought in court. Furthermore, disputes over land and mineral revenues in the Bakgatla area are modelled through group identity constructions and competing versions of custom.

The village level struggles and selected court judgements that are discussed in the next section demonstrate how disputes over control of mining revenues tend to underpin the power of local chiefs, particularly when the courts' interpretation of custom is rigid and distorted. The section below also demonstrates that the post-apartheid court judgements have ironically adopted the apartheid logic of interpretation of customary law which envisions all Africans as 'tribal' subjects whose property rights can be adequately controlled and mediated through the office of a chief.

Disputes, courts and custom

Kgosi Tidimane versus Jacob Pilane

Court battles between Bakgatla chiefs and their subjects predates the era of platinum revenues. During the rule of Kgosi Tidimane Pilane (*Kgosi* Nyalala's predecessor), there were some periods of resistance to the authority of the cheif. In one instance in 1953, Kgosi Tidimane imposed a levy of one ox per person on adult male members of his tribe for the purchase of the farms Middelkuil No 564 and Cyferkuil No 372. The combined price for both farms was £14 000 (21 044.10 US Dollar today).[54] Those who could not offer oxen were obliged to pay £15 (22.55 US Dollar) per person. A group of village residents, led by an activist Jacob Pilane (also a relative to the *Kgosi* Tidimane), filed a court petition in June 1956 accusing the chief of failing to account for the money he collected and 'wrongfully and unlawfully using and appropriating tribal funds for [his] personal benefit'.[55]

The hearing took place at the Transvaal Supreme Court in Pretoria on 28 June 1956. Jacob Pilane was listed as the only 'petitioner' against Kgosi Tidimane. Judge C. Bekker dismissed Jacob's application on 16 August 1956. His judgement was primarily based on the argument that the chief had no responsibility to account 'to anyone of his individual subjects' concerning the tribal accounts and that Jacob, although a member of the tribe, did not have *locus standi* to file a court application against the chief.[56] The judge continued:

> [I]n native law the chief, in circumstances such as the present is held accountable only to the tribe acting in, or through a *lekgotla* or tribal meeting ... the petitioner [Jacob], in his private capacity is not, in my view of the matter entitled to the relief he claims – reliefs personal to himself and not to the tribe.[57]

The judge also awarded costs against Jacob. This verdict was not the last of Jacob's troubles. The chief's loyalists in Moruleng harassed his family for challenging the chief and accused Jacob of trying to overthrow the chief. Since the judge awarded Kgosi Tidimane the costs in the case, this gave him more ammunition with which to punish Jacob. Jacob was unable to pay the legal costs, so Tidimane sent a group of men to his home to confiscate his cattle and agricultural tools by force. When this happened Jacob was in Swaziland, where he worked as a chef. One of his sons, who witnessed these events, said:

> The year was 1956 … when they came and took all my father's possessions. They came looking for my father's cattle. They took three cows together with all the ploughing equipment and left. They sold them to a white farmer called Piet Koos … in Pilanesberg.[58]

According to our respondents Jacob never recovered his confiscated property.

The judgement against Jacob Pilane relied on a distorted version of 'official' custom, which absolved chiefs from accounting to individual community members, thus providing them with enormous leverage to manipulate the downward accountability processes. As the only person entitled to call meetings (according to the 'official custom'), if a chief wants to avoid accountability he can simply refuse to convene community meetings. The four court disputes discussed below demonstrate how the courts still continue to use distorted 'official custom' in the post-apartheid democratic era. As platinum mining expanded in the Bakgatla area over the past decade, more and more community members display displeasure with *Kgosi* Nyalala Pilane's unilateral control over mining revenues. The chief has filed several court interdicts against a number of villagers who challenged his power over the Bakgatla-ba-Kgafela community.[59]

Kgosi Nyalala versus COBACO

In 2006, the regional court at Mogwase convicted *Kgosi* Nyalala and his close associate, Koos Motshegoe, on more than 40 counts of fraud and theft.[60] The fraud charges centred on the allegation that in 1998 *Kgosi* Nyalala signed three loan agreements to the value of R13 million (902,577 US Dollar, today) with the Land and Agricultural Bank of South Africa on behalf of the community, but without a community mandate. He pledged to repay this money through the annual royalties that the tribal authority receives from Anglo American Platinum. The regional court found that *Kgosi* Nyalala 'was not authorised to act on behalf of the tribe to enter into a loan agreement'.[61] Subsequently, the court denied the *kgosi* and his co-accused the right to appeal. *Kgosi* Nyalala's lawyer filed a petition with the then Judge President of the North West High Court, who in 2009 granted the chief and his co-accused permission to appeal against their criminal convictions.[62]

In September 2010, the high court upheld the application and acquitted *Kgosi* Nyalala Pilane and his co-accused of all criminal charges.[63] This ruling surprised and devastated the villagers. This blow was even more severe for members of the Concerned Bakgatla Anti-Corruption Organisation (COBACO), a village-based grass-roots movement which had worked hard, with limited resources, to get the chief convicted. It had taken them from 1997 to 2006 to finally get *Kgosi* Nyalala Pilane to court. One of the active members of COBACO explained:

> After that [*Kgosi*'s acquittal] we did nothing. We were there but we did not communicate, we did not hold meetings, things went quiet.[64]

Kgosi Nyalala versus Pheto and others

In May 2008, *Kgosi* Nyalala filed an urgent court interdict at the North West High Court against a group of residents led by David Pheto. Identifying themselves as the 'Royal House', Pheto and other disgruntled community leaders had called an urgent general community meeting (a *kgothakgothe*) in order to oppose the mining transactions that the chief was about to sign on behalf of the community. The meeting was to be held on 21 May 2008. The dissenting group of residents also wanted to pre-empt another general meeting called by the chief on 28 June 2008 to co-opt the community into endorsing a murky mining transaction. Through this meeting *Kgosi* Nyalala intended to obtain a tribal resolution for a transaction between Itereleng Bakgatla Mineral Resources (Pty) Ltd (IBMR) (owned by the Bakgatla-ba-Kgafela) and Barrick Platinum South Africa (Pty) Ltd (Barrick), a subsidiary of Barrick Gold Corporation.[65] The villagers opposed this transaction, mainly because they felt marginalised. They felt that the chief was unilaterally signing a mining contract that undermined their land rights without fully involving them.[66]

At the time, *Kgosi* Nyalala was facing a case of numerous instances of fraud and theft.[67] Pheto and other villagers demanded that *Kgosi* Nyalala step down from his position. In response, *Kgosi* Nyalala interdicted Pheto and five other leaders of the dissent 'from interfering with a … "general meeting"' which was to be held on 28 June 2008.[68]

In the North West High Court, Judge AM Kgoele consolidated the two interdicts and handed down the judgement, confirming both the interim interdicts by the chief against Pheto and others, on 3 December 2008.[69] The central argument in the judge's decision was that Pheto and five other community leaders did not have *locus standi* to call meetings of the tribe or to mobilise for the removal of *Kgosi* Nyalala from his position. Kgoele dismissed their claim that they were members of the 'Royal House' and refused them leave to appeal. The Supreme Court of Appeal also turned down their request for leave to appeal this decision[70].

Subsequent judgements at the North West High Court have reinforced Kgoele's decision. This has helped to suppress opposition against *Kgosi* Nyalala. For instance, in September 2011, Judge RD Hendricks confirmed an interdict by *Kgosi* Nyalala against Pheto and other leaders, preventing them from calling community meetings. In line with previous judgements and the North West High Court, the judge found that Pheto and others were not members of the 'Royal Family', therefore they did not have *locus standi* to call meetings or to represent any group of villagers in Bakgatla-ba-Kgafela territory.

The Jugde imposed punitive costs on Pheto and his fellow dissenters. He averred:

> It is quite apparent that the [r]espondents are doing everything within their means to unseat and undermine the authority of the [a]pplicants [*Kgosi* Nyalala and the Traditional Council] and to litigate as often as possible in an attempt to create confusion within the tribe. This behaviour borders on being vexatious. This, to my mind, calls for a punitive costs order.[71]

Village level disputes, whether over land, power or otherwise, are closely linked to struggles over mining revenues.

Kgosi Nyalala versus ba-Sefikile

In other cases involving local activists against *Kgosi* Nyalala Pilane, decisions at the North West High Court were no different. The court's decisions continue to endorse the version of custom that ossifies the chief's power over communal property and endorses the tribal authority as the only legitimate authority with *locus standi* to represent village residents. For instance, a group of residents, calling themselves the Bakgatla-ba-Sefikile Traditional Community Association (BBSTCA), claiming ownership of the mineral-rich farm Spitskop 410 JQ, sought a court order to compel the Minister of Rural Development and Land Affairs to transfer the ownership of the said farm to their names. This application sought to effectively remove *Kgosi* Nyalala's custodianship over this property and its mining revenues. The North Gauteng High Court transferred the application to the North West High Court on 31 January 2011. Judge Leeuw handed down her judgement on the case on 1 December 2011. She relied strongly on the perceived 'customary' custodianship of chiefs over communal land, and made reference to the Constitution and post-apartheid legislation that empowers chiefs to act as trustees over communal land. In this land dispute case, the judge dismissed the application of the BBSTCA with costs. The Judge, citing the Constitution and customary law, argued:

> In this matter I am enjoined by the Constitution to recognise that land that is held by the *Kgosi* or traditional leader on behalf of a tribal community should be dealt with in terms of legislations that have been enacted for the purpose of regulating amongst others, the ownership thereof as well as the role and powers of the traditional leaders.[72]

Kgosi Nyalala versus ba-Kautlwale

Another land claim by a group in Motlhabe village has produced a secession move which culminated in a protracted court case. The move to secede began with a notice in July 2009 from the leaders of land claimants calling themselves the Bakgatla-ba-Kautlwale to the traditional council of Bakgatla. Ba-Kautlwale, asserting strong claims over the mineral-rich farm Welgewaagd 515 (133 JQ) declared that Motlhabe village (located on Welgewaagd) was no longer under the jurisdiction and administrative control of the Bakgatla chieftaincy. In February 2010, *Kgosi* Nyalala's threats of legal action became real as soon as the leaders of the secession group circulated a meeting invitation. The invitation was addressed to 'The Residents of the Motlhabe Village' to discuss secession. *Kgosi* Nyalala tried to stop this meeting by filing an urgent court application to interdict the leaders. Through the interdict the chief prohibited ba-Kautlwale from convening any village meeting without the tribal council's permission. This action marked the beginning of a protracted legal battle between the chief and the ba-Kautlwale group. The battle went as far as the Constitutional Court. At the North West High Court on 30 June 2011, Judge Landman's judgement upheld *Kgosi* Nyalala's interdicts against the two ba-Kautlwale leaders and village activists Mmuti Pilane and Reuben Dintwe[73]. The judge argued:

> Any action by a parallel but unsanctioned structure that is neither recognised by law or custom seeking to perform or assume functions that are clearly the exclusive preserve of recognised authorities ought to incur the wrath of law[74].

The North West High Court and the Supreme Court of Appeal denied Pilane and Dintwe leave to appeal against this judgement. The lawyers who represented the two activists took the matter to the Constitutional Court which set aside the three interdicts in February 2013 mainly on the basis that these 'interdict adversely impact on the applicants' rights to freedom of expression, association and assembly'.[75] The Constitutional Court judgement was a landmark victory for traditional communities: it affirmed the freedom of expression, assembly and association of rural residents. It granted the Bakgatla-ba-Kautlwale a significant victory by removing all of *Kgosi* Nyalala's interdicts and it also reaffirmed the right of the villagers to meet without having to seek permission from the chief to do so.

Just a few weeks after the court case victory, the Bakgatla-ba-Kautlwale leaders filed an application to be recognised as an independent traditional authority in terms of the Traditional Leadership Governance and Framework Act of 2003. At the time of writing this article the outcome of this application was still pending.

Custom and 'community' divisions

Why are property disputes escalating in tandem with mining expansion? Why are the new rural struggles dominated by court battles? Is 'custom' so fragmented that it fractures the 'community'?

Rural struggles in the Bakgatla area epitomises the dilemmas raised by these questions. There have been several court disputes that coincide with the post-apartheid expansion of platinum mining in the Bakgatla area. The paper discusses five court disputes between the Bakgatla chiefs and ordinary villagers. Despite the fact that only one of the court cases discussed (*Kgosi* Tidimane versus Jacob Pilane) occurred during apartheid the interpretation of customary law in all five judgements significantly empowers the chiefs.

The distorted interpretation of custom by some courts in South Africa seems to shape the character of resistance in the Bakgatla area, particularly when such resistance is rooted on exclusive group claims over communal property. Such a phenomenon seems to resemble the character and function of power during the apartheid era. Mamdani's seminal work on 'the structure of power' during the late colonial period argues that the devolution of power to a local institution – a tribal institution – significantly enabled and empowered the state and local chiefs to crash rural resistance.[76] According to Mamdani, this institutional power was defended and legitimised through custom. As such, the institutions that enforced custom had authority to define it in the first place. The interpretation of custom by courts and their defence of institutional power epitomises this phenomenon. The 'new' rural customary order enforced by the post-apartheid state hinders the flexibility and consensus inherent in the living law. It is this inflexibility in the way the state and the courts conceive custom – as a rigid set of rules – which perpetuates inequality and disputes at village level. Rigid custom renders rural disputes zero-sum games. Groups or individuals that can prove in court that they have the legitimate version of custom are likely to win. This phenomenon destroys the natural consensus inherent in the 'living' African custom. Hence, I argue that the colonial crystallisation of custom made it difficult to become flexible when people's social conditions change, but easy to become fragmented (be subject to diverse, contested versions and interpretations) when disputes over property arise. Both the chiefs and ordinary villagers compete for the most 'legitimate' to bring forward during court disputes version of custom in order to have exclusive control over communal property. Mnwana argues that, by perpetuating a distorted interpretation of custom which protects and empowers chiefs the courts, not only leave the chiefs unaccountable, they also enhance marginalisation of ordinary rural residents, a phenomenon which makes residents lose faith in the justice system.[77]

The central feature of the village-level struggles that I have described in this paper are the diverse – and sometimes conflicting – group interests that run through them. This is in part explained by the fact that since rights over communal property are politically mediated by institutions of tribal authority, disputes over communal resources frequently take the form of power struggles at different levels in the tribal hierarchy. At the pan-tribal level, it was also seen that the broader struggle to remove the chief could be simultaneously undermined by conflicting group interests between and within different villages. This, I suggest, is particularly likely in a context where power at these lower political levels has the potential to translate into exclusive control over mining benefits. A critical effect is that contemporary struggles over land and mining rights contain within them political impulses that could ultimately lead to the disintegration of the 'tribal' community, as defined in the law. This is most graphically illustrated by the secession move in Motlhabe, as well as the assertion of political independence by the BBSTCA case in Sefikile. Indeed, rigid customary rules can be discriminatory and exclusive to other social categories.[78] In fact, Peters attests that the very negotiability and adaptability of customary property rights tend to favour the local elite (chiefs and other power holders) who use the rules of customary law to assert exclusive claims over communal property.[79]

Since rights over tribal property are politically mediated by chiefs, disputes over communal resources inevitably take the form of power struggles at different levels in the tribal hierarchy. Such a phenomenon is well illustrated in the work of Sara Geenen and Klara Claessens[80] in this issue, who argue that conflicts over land and resources in Congo are 'not solely about land and resources, but also about power and authority'. Chiefly custodianship goes against the existing character of 'communal' property rights. The state's empowerment of chiefs and the court's interpretation of custom tend to ignore the long local histories of resistance against 'tribal authorities' control over communal property.[81] In the Bakgatla villages, these struggles are mainly fought through competing versions of group land histories, custom and contrasting group identities. They function as social markers of rights to power and control over mineralised land and its revenues. The broader struggle to remove a powerful chief can be simultaneously undermined by conflicting group interests between and within different villages. This is likely to be the case particularly in contexts where power at these lower political levels also has the potential to translate into exclusive control over mining benefits. Therefore, it is possible to deduce that such escalating disputes in villages that are located on platinum-rich farms could point towards the fractious social relations within the community that are aggravated by continuous court battles that provide little room for natural consensus.

Postscript

While I was finalising this article in 2015, there were other significant court judgements that involved the chief and various land-claiming groups in the Bakgatla area. I have not included these recent judgements in this article. I discuss them in another contribution (still under preparation), where I expand on the argument raised in this article.

Disclosure statement

No potential conflict of interest was reported by the author.

Funding

This work was supported by the Human Rights and Governance Programme of the Ford Foundation through the Mining and Rural Transformation in Southern Africa (MARTISA) project at SWOP, University of the Witwatersrand [grant number 0120-6086].

Acknowledgements

I would like to thank the two anonymous reviewers for their helpful comments. I am also grateful to Gregory Maxaulane and Stanley Malindi, the two research interns with whom I conducted this study.

Notes

1. Comaroff and Comaroff, *Ethnicity, Inc.*; Manson, 'Mining and "Traditional Communities"'.
2. Peters, 'Inequality and Social Conflict'; Chimhowu and Woodhouse, 'Customary and Private Rights'; Mnwana, 'Mining and 'Community'.
3. Lund, *Local Politics*, 10.
4. Mamdani, *Citizen and Subject*, 110.
5. The term 'indigenous' here is applied loosely to refer to local African ('black') people in rural South Africa. I am aware that the term is highly contentious in other parts of the world.
6. Also called 'Bantustans': the ten ethnically defined semi-autonomous African 'states' that were created during apartheid for Africans/blacks in South Africa in order to exclude them economically, socially and politically.
7. Cousins, "'Communal' Tenure."
8. Mamdani, *Citizen and Subject*, 110.
9. Channock, "Neither Customary Nor Legal," 87.
10. Mamdani, *Citizen and Subject*, 109.
11. Delius, "Contested Terrain," 213.
12. Delius, "Contested Terrain," 213.
13. The Natives Land Act of 1913 and the Native Trust and Land Act of 1936.
14. Cousins, "'Communal" Tenure,' 6; Walker, 'We are Consoled', 6.
15. Delius, "Contested Terrain," 15.
16. Channock, 'Neither Customary Nor Legal'; Mamdani, *Citizen and Subject*.
17. Mamdani, *Citizen and Subject*, 109.
18. Capps, 'Tribal-landed Property'; Claassens and Gifillan, 'The Kalkfontein Land Purchase', 298.
19. Mamdani, *Citizens and Subject*, 109.
20. Mamdani, *Citizen and Subject*, 109.
21. Mamdani, *Citizen and Subject*, 109.
22. RSA, the Constitution of the Republic of South Africa No. 108 of 1996, Section 211. Pretoria, Government Printers.
23. The Constitution of the Republic of South Africa, Section 211 (3).
24. Claassens, "Customary Law and Zones of Chiefly Sovereignty," 362.
25. Bennett, 'Official vs Living Customary Law'.
26. Ibid., 138.
27. Comaroff and Roberts, *Rules and Processes*.

28. In 2014, South Africa produced 3.12 million troy ounces, followed by Russia (740,000 troy ounces), Zimbabwe (405,000 troy ounces), North America (400,000 troy ounces), Others (220,000 troy ounces) (Tatyana Shumsky, *Wall Street Journal*, http://www.wsj.com/articles/ accessed 25 June 2015).
29. The establishment of traditional authority areas in South Africa owes much to several pieces of colonial legislation, including the 'Native' Land Acts of 1913 and 1936 and the Bantu Authorities Act of 1951 (Act 68 of 1951, or BAA).
30. RSA, Traditional Leadership and Governance Framework Act, 2003 (Act 41 of 2003). Pretoria: Government Printers.
31. Claassens, "Resurgence of Tribal Levies," 14.
32. See Section 20 of the Traditional Leadership and Governance Framework Act of 2003 (Act 41 of 2003, or the TLGFA).
33. RSA, Communal Land Rights Act of 2004 (Act 11 of 2004) Pretoria: Government Printers.
34. Rural communities successfully opposed this Act on the basis that it undermined their private property rights enshrined in the Constitution and unfairly leveraged the powers of the traditional councils over vast areas of communal land. See Tongoane and Others v Minister for Agriculture and Land Affairs and Others (CCT 100/09) [2010] ZACC 10 (11 May 2010).
35. RSA, Traditional Courts Bill (B15 – 2008), published in Government Gazette No. 30902 of 27 March 2008). http://www.justice.gov.za/legislation/tradcourts/B15-2008.pdf (accessed 12 April 2012).
36. Mnisi, 'The Traditional Courts Bill', 3–10.
37. Details of these transactions are complex and beyond the scope of this article.
38. Mnwana, 'Are Communities Benefiting from Mining?' 30–33.
39. For details of the Marikana massacre see Botes and Tolsi, 'Marikana: One Year after the Massacre'.
40. Although important, mine labour issues are beyond the scope of this article.
41. Mnwana, 'Participation and Paradoxes'.
42. Claassens and Matlala, 'Platinum, Poverty and Princes in Post-apartheid South Africa', 116.
43. We have conducted 87 interviews with more than 100 respondents, so far.
44. Mnwana, 'Mining and "Community"'" 2015
45. See Binns et al., "Decentralising Poverty?" 28.
46. Capps, 'Victim of Its Own Success?' 71.
47. Capps, 'Victim of Its Own Success?' 72.
48. Mnwana, 'Mining and "Community"'.
49. Mnwana, 'Mining and "Community"'.
50. City Press, 'Madonsela to Probe Tribes' Lost Millions'.
51. These investors include Anglo American Platinum, Platmin Limited, Pallinghurst Resources Limited and the Industrial Development Corporation.
52. Khanyile, 'Bakgatla Tribe Men Facing Corruption Probe'.
54. Pretoria National Archives, PTD, Vol. 0, Ref. 1442/1956.
55. Ibid.
56. Mnwana, "Chief's Justice?"
57. Ibid.
58. S. Pilane, interview, Moruleng, 2 November 2013.
59. Mnwana, "Chief's Justice?"
60. National Director of Public Prosecutions v Pilane and Others (692/06) [2006] ZANWHC 68 (16 November 2006).
61. S v Pilane and Another (CA 59/2009) [2010] ZANWHC 20 (17 September 2010).
62. Ibid.
63. Ibid.
64. Interview, Mogwase, 19 July 2013.
65. Barric owned 10%, which it inherited when it acquired and integrated more than 80% of Placer Dome shares in 2006. The Bakgatla, through their holding company called IBMR, held 90% of the Sedibelo Project.
66. Interview, Lesetlheng, 7 August 2013.

67. As mentioned earlier, these are the charges that were brought to the Mogwase Magistrate's Court by the COBACO members.
68. Pilane & Another vs Pheto & 6 Others (1369/2008) [2008] ZANWHC (03 December 2008).
69. Ibid.
70. Mnwana, 'Chief's Justice?'
71. Pilane & Another vs Pheto & 5 Others (582/2011) (19 April 2012), para 55.
72. Bakhatla Basesfikile Community Development Association and Others v Bakgatla ba Kgafela Tribal Authority and Others (320/11) [2011] (1 December 2011), para 42.
73. Mnwana, 'Chief's Justice?'
74. Pilane & Another v Pilane & Another (263/2010), (30 June 2011), para 21.
75. Pilane & Another v Pilane & Another (CCT 46/12), (28 February 2013), para 70.
76. Mamdani, *Citizen and Subject*, 35–61.
77. Mnwana, 'Chief's Justice?'
78. Razayi, "Liberalisation and the Debates," 1488.
79. Peters, "Inequality and Social Conflict."
80. See Sara Geenen and Klara Claessens, *Different faces of access control in a Congolese gold mine*, In this issue.
81. Mnwana, "Mining and 'Community'."

Bibliography

Bennett, T. "Official vs Living Customary Law: Dilemmas of Description and Recognition." In *Land Power and Custom: Controversies Generated by South Africa's Communal Land Rights Act*, edited by A. Claassens and B. Cousins, 138–153. Cape Town: UCT Press, 2008.

Binns, T., G. Porter, E. Nel, and P. Kyei. "Decentralising Poverty? Reflections on the Experience of Decentralisation and the Capacity to Achieve Local Development in Ghana and South Africa." *Africa Insight* 35, no. 4 (2005): 21–31.

Botes, P., and N. Tolsi. "Marikana: One Year after the Massacre." *Mail & Guardian*. Accessed September 12, 2014. http://marikana.mg.co.za/

Capps, G. J. "Tribal-landed Property: The Political Economy of the Bafokeng chieftaincy, South Africa, 1837–1994." Unpublished Ph.D. Thesis, London School of Economics and Political Science (University of London), 2010.

Capps, G. "Victim of Its Own Success? The Platinum Mining Industry and the Apartheid Mineral Property System in South Africa's Political Transition." *Review of African Political Economy* 39, no. 131 (2012): 63–84.

City Press, 2012. "Madonsela to Probe Tribes' Lost Millions." Accessed August 22, 2014. http://www.citypress.co.za/news/madonsela-to-probe-tribes-lost-millions-20120211/

Chanock, M. *Law, Custom and Social Order: The Colonial Experience in Malawi and Zambia*. Cambridge: Cambridge University Press, 1985.

Chanock, M. "Neither Customary Nor Legal: African Customary Law in an Era of Family Law Reform." *International Journal of Law, Policy and the Family* 3, no. 1 (1989): 72–88.

Chimhowu, A., and P. Woodhouse. "Customary vs Private Property Rights? Dynamics and Trajectories of Vernacular Land Markets in Sub-Saharan Africa." *Journal of Agrarian Change* 6, no. 3 (2006): 346–371.

Claassens, A. "Customary Law and Zones of Chiefly Sovereignty: The Impact of Government Policy on Whose Voices Prevail in the Making and Changing of Customary Law." In *Land, Power & Custom: Controversies Generated by South Africa's Communal Land Rights Act*, edited by A. Claassens and B. Cousins, 355–382. Cape Town: UCT Press, 2008.

Claassens, A. "Power, Accountability and Apartheid Borders: The Impact of Recent Laws on Struggles over Land Rights." In *Land, Power & Custom: Controversies Generated by South Africa's Communal Land Rights Act*, edited by A. Claassens and B. Cousins, 262–292. Cape Town: UCT Press, 2008.

Claassens, A. "Resurgence of Tribal Levies: A Double Taxation for the Rural Poor." *South African Crime Quarterly* 35 (2011): 11–16.

Claassens, C., and D. Gilfillan. "The Kalkfontein Land Purchases: Eighty Years on and Still Struggling for Ownership." In *Land, Power and Custom: Controversies Generated by South Africa's Communal Land Rights Act*, edited by A. Claassens and B. Cousins, 295–314. Cape Town: UCT Press, 2008.

Claassens, A., and B. Matlala. "Platinum, Poverty and Princes in Post-Apartheid South Africa: New Laws, Old Repertoires." In *New South African Review 4*, edited by G. M. Khadiagala, P. Naidoo, D. Pillay, and R. Southall, 113–135. Johannesburg: Wits University Press, 2014.

Comaroff, J. L., and J. Comaroff. *Ethnicity, Inc. the Zulu Kingdom Awaits You*. Chicago: University of Chicago Press, 2009.

Comaroff, J. L., and S. Roberts. *Rules and Processes: The Cultural Logic of Dispute in an African Context*. Chicago: University of Chicago Press, 1981.

Cousins, B. "Characterising 'Communal' Tenure: Nested Systems and Flexible Boundaries." In *Land, Power and Custom: Controversies Generated by South Africa's Communal Land Rights Act*, edited by A. Claassens and B. Cousins, 109–137. Cape Town: UCT Press, 2008.

Delius, P. "Contested Terrain: Land Rights and Chiefly Power in Historical Perspective." In *Land, Power and Custom: Controversies Generated by South Africa's Communal Land Rights Act*, edited by A. Claassens and B. Cousins, 211–237. Cape Town: UCT Press, 2008.

Khanyile, G. "Bakgatla Tribe Men Facing Corruption Probe." *iol News*, October 14, 2012. Accessed July 12, 2013. http://www.iol.co.za/news/south-africa/north-west/bakgatla-tribe-men-facing-corruption-probe-1.1402519#.VCA4PPmSy8A

Lund, C. *Local Politics and the Dynamics of Property in Africa*. Cambridge: Cambridge University Press, 2008.

Mamdani, M. *Citizen and Subject: Contemporary Africa and the Legacy of Late Colonialism*. Princeton: University Press, 1996.

Manson, A. "Mining and 'Traditional Communities' in South Africa's 'Platinum Belt': Contestations over Land, Leadership and Assets in North-West Province c. 1996–2012." *Journal of Southern African Studies* 39, no. 2 (2013): 409–423.

Mnisi, S. "The Traditional Courts Bill: Controversy around Process, Substance and Implications." *South African Crime Quarterly* 35 (2011): 3–10.

Mnwana, S. "Participation and Paradoxes: Community Control of Mineral Wealth in South Africa's Royal Bafokeng and Bakgatla Ba Kgafela Communities." Ph.D. Thesis, University of Fort Hare, 2012.

Mnwana, S. "Are Communities Benefiting from Mining? Bafokeng and Bakgatla Cases." *South African Labour Bulletin* 37, no. 3 (2013): 30–33.

Mnwana, S. "Mining, Accountability and the Law in the Bakgatla-Ba-Kgafela Traditional Authority Area." *South African Crime Quarterly* 49 (2014): 21–29.

Mnwana, S. "Mining and 'Community' Struggles on the Platinum Belt: A Case of Sefikile Village in the North West Province, South Africa." *The Extractive Industries and Society* 2 (2015): 500–508.

Peters, P. E. "Inequality and Social Conflict over Land in Africa." *Journal of Agrarian Change* 4, no. 3 (2004): 269–314.

Razavi, S. "Liberalisation and the Debates on Women's Access to Land." *Third World Quarterly* 28, no. 8 (2007): 1479–1500.

Walker, C. "'We Are Consoled': Reconstructing Cremin." *South African Historical Journal* 51, no. 1 (2004): 199–223.

Disputes over gold mining and dispossession of local afrodescendant communities from the Alto Cauca, Colombia

Irene Vélez-Torres

ABSTRACT
In 2001, the local afrodescendant community from La Toma faced a double threat. First, new mining legislation was introduced and their ancestral mining activities were declared illegal. Second, paramilitary groups entered the territory and initiated a regime of terror that continues to date. Mining titles were conceded to multinationals and private parties, while the communities were restricted in their access to land-based resources. This paper deals with different dimensions of local dispossession resulting from violent mechanisms as well as from the neoliberal adjustment of the mining regulatory system in Colombia.

Introducing Colombian extractivist geographies

The year 2001 is deeply ingrained in the memory of the afrodescendant community from La Toma, in the Alto Cauca region of Colombia.[1] In April, the most important paramilitary group at the time, the Autodefensas Unidas de Colombia (AUC), took control of the region and began a regime of terror that resulted in selective murders, forced displacement and confinement, and violent restrictions to the communities' management of natural resources. In the Alto Cauca, the AUC commander Ever Veloza (alias H.H.) has confessed to several crimes, including the massacre of more than 40 indigenous people from the Naya region in 2001[2] and the displacement of 100s between 2001 and 2004. Though the AUC was dissolved in 2004 after a process of 'social reinsertion', communities from the region have confirmed that paramilitary organisations continue to exist under new names.

During 2001, a new Mining Code was promulgated and the ancestral gold mining activities of the afrodescendant community were made illegal until such time as a mining title was allocated by the national government. In subsequent years, under protection of AUC paramilitaries, backhoes entered the region and started the uncontrolled extraction of gold along the riverbanks that flank the Cauca river. The local community was never informed about the legal process to formalise their mining activities, and the arrival of death threats and increasing violence in the region did not render circumstances fit to engage in lengthy bureaucratic procedures. Three years later, in 2004, the local inhabitants came to know that several mining titles had been allocated to multinationals and private actors. According to the legal rights acquired by the entitled parties, local afrodescendants had to evacuate their ancestral territory as they were considered illegal inhabitants, and their mining an illegal economic activity.

The complexity of mining practices and actors in La Toma can be better pictured if categorised according to the different actors, their extractive practices and legality:

(1) Ancestral mining: is mining developed by afrodescendants, traditional inhabitants of the Alto Cauca, legally recognised by Law 70/1993 and by the Auto 005 of 2009. The community in La Toma does not use environmentally damaging chemicals such as mercury or cyanide in their mining. Ancestral mining is based on the 'barequeo' technique, where miners use a wooden plate to separate the gold from other material; sometimes using other techniques such as water pumps and mechanical rock-crushing machines. In the case of La Toma, this form of mining has not been legalised via a formal mining title and, therefore, it remains illegal from the perspective of the Mining Code.

(2) Mechanised small-scale mining or backhoe mining: in La Toma, this mining is developed by actors who are foreign to the community, often guarded by illegal paramilitary forces such as Rastrojos or Águilas Negras. At times, the guerrillas Fuerzas Armadas Revolucionarias de Colombia (FARC) have also participated in this form of mining through informal taxation of the backhoe mining or through the provision of paid protection for the mining operations. Next to the backhoes, this type of mining uses mercury and cyanide to separate gold from other materials. This mining has not been granted a mining title or an environmental license; thus, from the perspective of the Mining Code and the public institutions for environmental control, it is an illegal economic activity. Whether due to lack of political will or lack of institutional capacity, local and national authorities have had great difficulty controlling the influx and activity of illegal mechanised mining.

(3) Large-scale mining: in La Toma, this type of mining has been granted legal permits but extraction has not yet started. This contention can be understood as the result of resistance by the local population and supporting social organisations. According to the Mining Code, this mining is characterised by its productivity, efficiency and technical capacity, standards that are more often acquired by multinational corporations. As shown by the Observatorio de Discriminación Racial (2011, 35), in the municipality of Suarez where La Toma is located, the most extensive mining licences have been granted as follows: 1717 ha to Andrés Rendle, 1404 ha to Multinacional Anglo Gold Ashanti, 314 ha to Raúl Fernando Ruiz and 99.7 ha to Héctor Jesús Sarria. Although this titling is legal as it is based on a formal process regulated by the national government, it does not necessarily receive a 'social license'.[3]

Zooming out from the local mining settings of La Toma, it is important to analyse how different mechanisms to promote extractive industries have led to profound transformations of landscapes across Latin America. Numerous communities have been disenfranchised from their lands and environmental assets, generating local violence, displacement and radical change to their traditional livelihoods. Since the early 1970s, many scholars criticising these impacts have identified the economic dependency on minerals, petroleum and other extractive industries as a curse for Latin America and other tropical regions.[4] The 'resource curse' argument emphasises the undesired stream of power relations that dependency produces and reproduces, which overlooks citizenship rights and create underdevelopment rather than the promised modernisation, progress and development.[5]

Neoliberalism has impacted on legislative and institutional designs in relation to mining since the early 1990s, and is expressed in the promotion of foreign direct investment, the change of regulations, and the consequent limitation of state control over the exploitation of mineral assets. From a political economic perspective, the expansion of mining frontiers after the 1990s is the result of a spatial fix of capital to resolve the crisis of overaccumulation, where new geographies are intended to be included in the production–distribution–consumption process in order to access new resources as well as new consumer societies[6]. What follows is a process referred to as extractivist: capital flows accumulate outside of the site of extraction, and damages and risks increase for the local communities.[7]

In the early 2000s, a leftist turn in the politics of several governments in Latin America prompted new hope regarding the management of natural resources in the region. New political constitutions in Ecuador (2008) and Bolivia (2009); greater taxes and royalties in Bolivia, Ecuador and Venezuela; the re-nationalisation of several mining and energy companies in Argentina and Bolivia; and state funding of resource exploitation in Brazil were some of the signs of change.[8] Although revenues have been funnelled to social programs and profit from mining capital was not solely accumulated by the same private hands, the leftist turn has not proved different in its model of the exploitation of nature. On the contrary, some academics have argued that neo-extractivism has continued to open new geographies to resource exploitation.[9]

While the so-called Progressist (and leftist) turn has not been able to shift the dependency on raw materials, conventional neoliberal mining has been reinforced in Colombia, among other countries in the region.[10] In the year 2001, the Colombian government designed a new flexible and foreign investment-friendly mining code to strengthen the corporate extractive mining industry. As a result of this legislation, foreign mining investment increased from 627 million USD in 2003 to 2162 million USD in 2011.[11] In the same period, the mining contribution to the GDP enlarged from 1.8% in the year 2002, to 14.5% in 2011.[12] The number of mining concessions has shown a similar exponential increase from 2952 titles in 2004, to 8905 in the year 2010.[13] Also, concessions have shown an exponential increase in the total entitled area (see Table 1). On a regional scale, a study of the mining sector has shown that 241 mining titles have been granted and 350.447 ha have been titled in the Cauca department; by the year 2012, more than 1,106,665 new hectares were waiting for legal allocation.[14]

The mining boom has not only been reflected in national legislation and official national figures. It is also shown in the illegal ventures of miners who, guarded by illegal armies, use mechanised backhoes to excavate in the watersheds of strategic biomes such as the Amazon, the Pacific and the inter-Andean valley. While new extractivist geographies have been created, conflicts have multiplied. The term extractivist geographies refers to the uneven intersection between economy, ecology and society when an economic activity accumulates profit in places and by people different from the settings of resource-extraction, and results in direct or indirect impacts on local socio-ecosystems affected by contamination, degradation and other processes. Recent research by national NGOs has reported more than 274 collective actions opposing mining and oil projects between 2001 and 2011.[15] Another study developed by the Environmental Justice Organizations, Liabilities and Trade project in 2014

Table 1. Mining entitlement in Colombia 2004–2010.

Year	2004	2005	2006	2007	2008	2009	2010
Number of mining titles	2952	3459	4283	5580	6603	8267	8905
Number of entitled hectares	1,040,323	1,202,915	1,569,001	2,416,199	3,418,893	4,385,886	4,805,940
Entitled national territory	0.9%	1.1%	1.4%	2.1%	3%	3.8%	4.2%

Source: Based on 'Censo minero departamental colombiano' (Misterio de Minas y Energía, 2010).

has shown that more than half of the 72 socio-environmental conflicts in the Andean region occurred as a consequence of the mining expansion.[16]

To understand the tension between legal and illegal mechanisms creating extractivist geographies in Colombia, we can benefit from the concept of dispossession as discussed by David Harvey[17]. According to his reflection on Rosa Luxemburg's explanation of primitive accumulation, capitalism is an open system that permanently needs an outside(r) to stabilise itself: for capitalism to expand it is necessary to incorporate cheaper inputs and wider markets. According to Luxemburg, as read by Harvey, capitalism satisfies these needs by forcing into the system new territories for extraction and new people for consumption. Harvey explains that, in order to incorporate new territories and consumers, two prominent mechanisms are utilised: first, extra-economic force (such as usurpation, spoliation, expropriation, fraudulent appropriation, robbery and fraudulent alienation); and second, legal and normative deprivation of local communities. In what follows, these mechanisms of dispossession will be used as analytical tools to examine (i) the legal mechanisms that have given a formal structure to extractivist geographies in Colombia, and (ii) the coercive practices that coexist and are exercised locally, and are meant to protect legal and illegal private accumulation.

First, I present the activist and participatory methodology applied in the research. Second, I discuss the national mining code, and the ethnic mining rights recognised in it. I then present the case of La Toma in order to describe how afrodescendants from the Alto Cauca have confronted a double territorial pressure of dispossession: by the legislation that has benefited private mining entitlement, and by the armed groups that have guarded the illegal mining of backhoes. Finally, I provide some conclusions on how the legal framework in relation to mining has, along with territorial violence, been used to dispossess local ethnic communities in Colombia.

Methodology

This paper discusses some of the results of research into conflicts over land and water in Colombia and Brazil.[18] In Colombia, the research focused on the inter-Andean valley of the Cauca river, in a region known as the Alto Cauca in the south-west of the country. It followed a longitudinal methodology where many long and short visits have been made to the territory between 2009 and 2016.

The epistemological perspective in the investigation was to research *with* and *for*, rather than *on*, the social movement.[19] Accordingly, this research aimed at contributing to the local contestation to extractivism by critically analysing the struggles over land-based resources. A Participatory Action Research (PAR) approach was embraced from an activist perspective, recognising that knowledge is a situated process with the participation of different actors over time: local communities, academics, activists and public officers.[20] The collective crafting of knowledge has made it a communal good, shared and protected by the community. It is understood, therefore, that it has not been due to this particular research that social organisation and contestation has emerged, but rather this research aimed at contributing to the process of resistance.

Drawing on this perspective, the research aims and objectives were prior agreed upon with different Community Councils[21] from the Alto Cauca region, and with the Community Council from La Toma in particular. An explicit openness about political ideas and engagement with the social organisation was part of the ethical framework of the research. As the research agenda overlapped with the social movement's agenda, leaders and people from the local community acquired confidence towards the work, the use of the information and the long-term commitment to their struggles.

Formal research methods have included social mapping, participatory documentaries, archival research, workshops and interviews.[22] In 2012, I lived in a mining site for eight months, and I later visited the region frequently during periods of crisis upon request of local actors who asked me to be present. I have participated in demonstrations, facilitated dialogues between leaders and local, regional and national governments, made public press releases to support the communities' defence of their territories, and served as facilitator in the communities' dialogue with national governmental institutions and international human rights organisations.

The normative expansion of the Colombian mining frontier

Addressing the Colombian legal framework on mining is crucial for understanding local conflicts over gold mining in the Alto Cauca. The following section provides an introduction to the Mining Code, and discusses some of the opportunities and restrictions that it brings to ethnic communities in terms of accessing mining resources.

The Mining Code was legislated in Law 685/01. Given the fact that mining regulation was a top priority to the government at that time, the Code previewed the formalisation of traditional mining by establishing Special Reserves (Article 31). These zones were meant to be granted to miners who could demonstrate that they had mined in a particular mining setting before 2001. This right is not restricted to ethnic communities and, it does not reflect, nor protect the cultural particularities of indigenous and afrodescendants. To overcome this limitation, and in coherence with the protection of ethnic communities condensed in the Political Constitution of 1991, the Mining Code recognised the possible allocation of Mining Zones to ethnic communities (Articles 121–136). However, even though the Code grants indigenous persons and afrodescendants the right to prior entitlement (Articles 124 and 133 respectively), it does not provide the communities the right to veto. In other words, even communities whose livelihoods do not include mining activities are trapped in a mining curse: when not displaced by the multinational capital, local inhabitants are forced to become miners, either under their own management through the priority right, or through exploitation by an actor foreign to the community.

Implications of the mining legislation do not stop with the forced creation of extractivist geographies. A community that has traditionally worked on mining, for example, and decides to make use of its right to prior allocation, is expected to meet certain requirements of productivity, efficiency and technical capacity that make it very difficult to uphold the entitlement. As a consequence, economic standards do not only imply a forced change to increase productivity through changing technology, but also the search for external financial support to meet the requirements. The current mining legislation implies, therefore, a normative-forced transition from subsistence economy to an economic system that, based on accumulation of surplus, implicitly pulls (i) new private actors and (ii) foreign financial capital to the mining settings.

Recently, a network of NGOs has researched cases of indigenous communities from the Chocó region, who have demonstrated limited capacity to maintain their rights to ethnic mining entitlement.[23] According to the findings, even when a community is granted an Indigenous Mining Zone, the requirements on productivity make it difficult for the local miners to reach the standards. In addition, the communities are exposed to threats, murder, assaults and other forms of violence that generate coercive pressure over their territory and economic practices. Despite the mentioned difficulties, the right to prior entitlement is an aspiration of most ethnic mining communities as it is the only scenario they see through which they can continue to control their territory.

Against the aspiration to ethnic entitlement, a major difficulty arises regarding mining by afrodescendants in the Alto Cauca: the right to prior mining has been made dependent on collective land titling by the state land agency, INCODER[24] (Article 131 of the Mining Code). Although it sounds reasonable, this requirement constitutes a bureaucratic barrier to afrodescendants whose lands have not been formalised since Law 70 was formalised in 1993. While an unprecedented process of formalisation of collective land ownership has taken place between 1996 and 2015 in the Pacific coast[25], there have not been any 'collective territories' granted title in the inter-Andean valley within the framework of the Law 70. As a result, ethnic mining entitlements have been restricted to communities from the Pacific coast, and ancestral mining communities from the inter-Andean valley have been excluded from such rights. These bureaucratic barriers constitute a primary reason that communities from this region have not been able to organise extractive projects that promise social, economic and ecologic sustainability.[26]

La Toma: a mining site of dispossession

Besides legislation, the Colombian mining boom has been expressed through the actions of illegal backhoes that, with limited institutional control, extract gold from the river basins of important ecosystems such as the Amazon, the Pacific and inter-Andean valley. In the case of this last region, legal and illegal mining have contributed in different ways to local dispossession and socio-environmental conflicts. Even though more than 30 backhoes were seized after illegal gold extraction from the region in 2014, the proliferation of the backhoes has not been efficiently controlled by public institutions.[27] A relevant case in the government's struggle – and failure – to control illegal mining have been the threats to Ricardo Cifuentes, government official in Santander de Quiliachao, who was forced to leave his job and the region in response to the threats to his life after denouncing the action of different backhoes in a mining site known as San Antonio.[28]

In the region, a salient conflict over the access and control of gold mining has been the one in La Toma District, which is home to more than 1300 households of afrodescendants. The first inhabitants of La Toma were brought as slaves to the local gold mines. After the abolition of slavery in 1851, many afrodescendants developed diverse livelihoods based on agriculture, fishing and artisanal gold mining. Despite their long history of gold extraction, local inhabitants have disputed the control over their traditional mining practices and territory since 2001, against the interests of foreign mining corporations, illegal backhoe miners and the government. By defending ancestral small-scale gold mining, communities are opposing the pressure of the normative system – that has led to private mining titles – and are confronting the pressure of violence by illegal armed groups that have guarded both legal mining operations and illegal mining backhoes.

In the year 2000, the Ministry of Mines conceded the first mining title in La Toma to a private person, who aimed at selling the 99.7 ha allocated to him to the multinational corporation Anglo Gold Ashanti. By the year 2009, more than 95% of the 7000 ha that make up the district of La Toma, was granted in concession by the national government to private actors.[29] From the community's perspective, such entitlement was illegitimate as it was conceded during a period of regional control by the AUC paramilitaries, who violently restricted the community's autonomy to decide over their land-based resources. Controversially, as the mining territory was privatised via mining concessions, local artisanal miners were automatically declared illegal miners. The illegality of their mining was supported in the Mining Code that states that only formal title allocation makes mining legal.

As was set out in the previous section, and though the Mining Code allows Mining Zones for ethnic communities, the right to formalise ancestral mining has been made conditional upon the previous concession of collective territories to afrodescendant communities by the INCODER. In the case of La Toma, such formalisation of collective land ownership has not yet been granted due to the complex bureaucracy and a correlated lack of political will by local and national government. Ancestral miners have been rendered illegal as a result of a delay on the collective land entitlement. Such crafted bureaucracy resonates with the argument by Campbell in this collection. According to her research in the African context, neoliberal mining reforms have introduced normative categories that are uprooted from reality, creating a dichotomy between regulatory frameworks and empirical realities.[30]

Local miners respond to being categorised as 'illegal' by the authorities with indignation. Expressing the collective feeling, one inhabitant from La Toma declared:

> One thing has to be said in relation to the concept 'illegal'... because the Mining Code has placed us to be the 'illegal' whereas, in reality, the ones that are illegal are the multinationals (...) I want to formulate a question: is it us, the ones that have done traditional mining for centuries in this country, the ones who are illegal miners? For me, the illegals are the ones that make big environmental impacts across the country and only generate pollution. (Speech in the Encuentro Nacional de Consejos Comunitarion, Bogota 05/2012).

This reflection, directed to more than 1000 participants of a national congress of afrodescendant leaders in the year 2012, invites a critique on how legality does not imply legitimacy. As a matter of fact, legality can contradict what is considered legitimate when, as in the case of La Toma, traditional inhabitants are turned into illegal mining actors on their own land, even though they are practicing the same ancestral activity that the national constitution is supposed to protect. Making resourceful use of the options given to traditional miners by the Mining Code, in September 2015 the Mining Cooperative from Suarez, along with other social organisations, demanded that the Mining Ministry formalise traditional mining in the region under the legal concept of Special Reserves. However, this aim raised new difficulties: the ministry pushed miners to accept individual formalisation of the mining extraction, while communities and social organisations opposed such an approach because it did not reflect the collectiveness performed in mining and it exposed local inhabitants to violent pressures from private parties who may have been interested in making use of the formalisation granted individually.

Similar pressures have indeed been experienced by local miners who have confronted the violent entry of backhoes into the territory (particularly the river 'Ovejas') since 1997. Guarded by illegal armed groups, the 'retreros', named after their ownership and operation of backhoes (*retroexcavadoras* in Spanish), have made it very challenging for the local afro-descendants to exercise territorial control. A local leader said it clearly:

> Mining and territorial governance is complex because one can say to the foreign miners 'you do not enter here' but in practice one must recognize that the armed conflict, and what the paramilitaries are doing through their politics of fear and terror, is actually implementing by force a local policy of extraction. (Interview with a leader from the *Consejo Comunitario* of La Toma, Bogota 05/2012)

The incursion of illegal backhoes, the granting of mining permits to multinationals and private parties, and the legal framework to govern the subsoil that has restricted access to land and gold by local afrodescendants, have resulted in the unequal allocation of, and access to land-based resources in the Alto Cauca. The subsequent dispossession of the local community has different variants: (i) dispossession of land by the mining titles granted to private actors; (ii) dispossession of land-based resources by the incursion of backhoes that have reduced the number and quality of places for fishing and artisanal mining on the Ovejas river bank;[31] (iii) dispossession of collective land uses and cultural practices resulting from the postponement of collective land entitlement and collective mining licensing; and (iv) dispossession of the people's ability to control land and land-based resources due to the presence of illegal armies, and the threats to, and murders of community members that have come with it.

Conclusions

The analysis of the ethnic component of the Mining Code has exposed some of the practical difficulties in the normative regime to incorporating traditional extractive practices of afro-descendant mining communities. First, the options given to ethnic communities that are settled in mining-rich territories are to become legal miners or accept that mining titles be granted to foreign parties; alternatives to mining are not considered, even when traditional livelihoods are not centred around mining. As a result, formalisation runs the risk of creating new extractivist geographies and mining communities, while it can change the equilibrium of traditional livelihoods characterised by diversity. In other words, within this juridical context, communities with diversified economies may be forced to commence or intensify mining, displacing economic activities that have traditionally been complementary to mining. Second, formal pressures of productivity and efficiency have created the risk of forcing a transition in the local economies and landscapes that may have major impacts on the traditional practices and worldviews of ethnic communities. It is not the wealth in underground resources, then, that curses the local communities. Political decisions are being taken, that result in the community having to face restrictions on their autonomy in ancestral territories and traditional practices.

In the case of afrodescendant communities from the Alto Cauca, the situation has an additional variant: collective land-titling, granted by the Law 70 of 1993, has been delayed for more than 20 years. Until collective land formalisation is granted to afrodescendant communities in the region, they are legally disqualified as subjects of prior mining entitlement and, consequently, are more exposed to the possible impacts of mining by multinational and private actors. Even though local communities in the Alto Cauca are ancestral miners, the on-paper opportunity given by the Mining Code to formalise their extraction of gold has not granted real access to mining resources. As a result of the denial of their rights to land and gold, communities in the Alto Cauca are forced to face a double bureaucracy: one, to formalise their collective land ownership; and two, to formalise their mining activities.

Alternatively, local miners have attempted a different form of mining formalisation through a Special Reserve. Though this legal resource has been used by the local community to limit mining allocation to private parties, this concept has proven insufficient to acknowledge the collective practices structuring the mining tradition of ethnic communities in the region. The several difficulties found in the legal system to legalise local ancestral mining and to stop the government's mining allocation to multinational and private parties, have resulted in the need to differentiate what is legal from what is legitimate. While legality has been experienced as a mechanism of dispossession, the concept of legitimacy can be framed as an oppositional discourse by local communities to defend their culture, territory and ethnic rights.

Besides the juridical system that limits access to land and mining by afrodescendant communities in the Alto Cauca, another prominent mechanism used to control land-based resources has been the para-State military violence that has paralleled and defended legal mining entitlement, and guarded illegal mining extraction. The uneven co-relation has been denounced by the community from La Toma, who has characterised the mining entitlement as illegitimate because it took place while the community was victimised by the AUC paramilitary group. Conscious of the social and ecological damage, local communities have also requested local authorities to stop illegal gold extraction by the backhoes.

Dispossession in the Alto Cauca has occurred through a combination of normative and violent mechanisms. While the illegal actions of the backhoes are expanding the mining frontier beyond State control, dispossession has also resulted from the norms that aim to enhance foreign corporate capital and large-scale mining projects. Accumulation by dispossession, in the case described here, has not only been shown to be coercive, but also legalised through a sophisticated structure of norms that has restricted the communities' autonomy to decide over their resources and territory. This paper has argued that normative and coercive mechanisms have been complementary in dispossessing ancestral miners from their right to collective land, to land-based resources, to their traditional mining culture and to exercise control over their territory.

By addressing the complexity of extractivist geographies in Colombia, this article contributes to our understanding of the internal armed conflict in two ways: first, it helps to clarify how the armed conflict has occurred in practice, what interests have driven it, and which actors have been responsible. Secondly, it provides empirical support to the official recognition of the existence of subjacent factors to the internal armed conflict in the case of afrodescendant communities whenever a legal or illegal economic activity has (i) been supported by an illegal army and (ii) has affected the human rights of an afrodescendant community (based on the Decree 4635 of 2011, complementary to the Victim's Law). In the context of the current peace dialogues, it can be argued that an assertive transition to 'peace' may only be achieved once the conflicts over mining and over other land-based resources are not only accepted as structural to the historical internal armed conflict, but also addressed adequately by the State.[32] Consequently, social participation should be seriously considered while effective mechanisms to grant ethnic rights should be established, ones that reflect clear political and institutional will, as well as sufficient financial capacity.

Disclosure statement

No potential conflict of interest was reported by the author.

Funding

This work was supported by the NWO–CoCooN.

Notes

1. Afrodescendants, Afro-Colombian or black communities are considered in Colombia ethnic groups that, next to indigenous and Rom communities, have been granted differentiated ethnic rights. According to the Political Constitution of 1991, the nation was recognised as culturally diverse and the State was given the legal responsibility to protect ethnic communities and their territories. As declared in Law 70/1993, among other ethnic rights granted to afrodescendants, the right to *collective territory* is prominent as territory is considered a necessary space for the cultural reproduction and the well-being of these groups. To read about the ways afrodescendants have self-identify and socially organise since 1970 to claim rights such as territory, culture and autonomy, see Restrepo, *Etnización de la negridad*.
2. Ethnographic research on the paramilitary massacre and displacement in this region is presented in Jimeno et al., "Experiencias de violencia".
3. This concept implies that the mining entitlement is not legitimate in the eyes of the local community.
4. A historical analysis of dependency in the Latin American can be found in Mignolo, "Philosophy and the Colonial"; and Acosta, "Extractivism and Neoextractivism".

5. To examine dependency from the perspective of citizenship, see Nem Singh, "Reconstituting the Neostructuralist State".
6. More on the behaviour of mining in the neoliberal context: Gordon and Webber, "Imperialism and Resistance," 63–87; and Bridge, "Mapping the bonanza," 406–21.
7. To approach the debate on extractivism, see Gudynas, "Transitions to Post-extractivism"; and Martinez Alier, "Los conflictos ecológico-distributivos".
8. Gudynas, "La ecología política del progresismo," 147–67.
9. An interesting analysis of the limits encountered by leftist governments to divert from neoliberalism can be accessed in Andreucci and Radhuber, "Limits to 'counter-neoliberal' Reform". Also, on the governmental challenges to deal with extractive industries in Latin America, see Gudynas, "Transitions to Post-extractivism".
10. A comparative analysis of extractivism in Colombia and Peru has been presented by Velez-Torres and Ruiz, "Extractivismo neoliberal minero".
11. Ministerio de Minas y Energía, "Censo minero departamental."
12. The following video presents one of the dialogue scenarios created between stakeholders: https://www.youtube.com/watch?v=Tlwm8nLQNmc&feature=youtu.be.
13. El nuevo siglo, "Crece aporte de minería al PIB."
14. For the numbers of titles and hectares allocated to private and multinational mining in Colombia, see Duarte, *Situacion minera en los territorios campesinos*.
15. CINEP & PPP, *Minería, conflictos sociales*.
16. Perez, "Conflictos ambientales en Colombia."
17. While Chapter 27 of Capital by Marx provides a sharp explanation on accumulation by dispossession in the context of Primitive Accumulation, I would like to focus on the interpretation made by David Harvey in *The New Imperialism*.
18. The project, financed by NWO–CoCooN, took place between 2010 and 2015. See: http://www.landsandrights.blog.com/.
19. On different relationships with social movements in the context of research, see Ballard et al., "Globalization, Marginalization and Contemporary Movements", 615–34.
20. Some of the researchers who have discussed an activist, organic, participative and collaborative research are Fals, "Investigating Reality in Order to Transform it", 33–55; Hale, "Activist Research", 96–120; Katz, "Vagabond Capitalism", 709–28; and Harding, "Rethinking Standpoint Epistemology", 218–29.
21. The Community Council (Consejo Comunitario in Spanish) is the figure created in 1993 by the Law 70 to formalise the traditional, social and political organisation of Afro-Colombian communities. Councils gather yearly in assemblies to take major decisions for the community and the territory, and are administrated through a board of directors elected every year.
22. Velez-Torres, "Reflections on a Participatory Documentary Process", 299–306; and Velez-Torres et al., "Cartografía social", 59–73.
23. ABC Colombia, CINEP, and Tierra Digna, *Alimentando el conflicto en Colombia*.
24. INCODER is the public institution responsible for formalising land ownership.
25. On the land titling process in the Pacific region, see Offen, "The Territorial Turn"; and Restrepo, *Etnización de la negridad*.
26. An example of alternative mining projects that are meant to be more sustainable and socially equitable is Oro Verde in the Chocó region. For more information, go to the GOMIAM project in: http://www.gomiam.org.
27. Several media have reported drawing from critical research on the environmental impact of illegal mining. See: El espectador, "Ejército destruye seis retroexcavadoras"; Proclama del cauca, "Incautadas máquinas retroexcavadoras"; and Tenthoff, "Siguiendo la ruta de la minería"; Semana, La guerra contra la nueva coca.
28. Vanguardia.com, "Defensoría llamó la atención."
29. On local mining entitlement: Observatorio de discrimacion racial, *La disputa por los recursos*.
30. Campbell, "Distancing from an Investment-led Perspective".

31. To address the discussion on dispossession of land-based resources in terms of quantity and quality, see Bebbington et al., "Mining and Social Movements", 2888–905; and Bebbington, "Underground Political Ecologies".
32. As an illustration of the connection between peace and land-based conflicts, afrodescendant communities agreed on not going to La Habana, where the talks with the FARC are talking place, arguing not only that the dialogue has gone too far without their participation, but that the government has failed in accomplishing old and recent agreements to grant the rights of ethnic communities. See Elmundo.com, "Afrodescendientes no acompañarán la mesa."

Bibliography

ABC Colombia, CINEP, and Tierra Digna. *Alimentando el conflicto en Colombia: El impacto de la minería de oro en Chocó* [Feeding the conflict in Colombia: The impact of gold mining in Choco]. Bogotá: ABC Colombia, 2012. http://www.abcolombia.org.uk/downloads/ABColombia-Choco_mineria_SPANISH.pdf (revised on the February 28, 2016).

Acosta, Alberto. "Extractivism and Neoextractivism: Two Sides of the Same Curse." In *Beyond Development. Alternative Visions from Latin America*, edited by Lang Miriam and Dunia Mokrani. Amsterdam: Fundación Rosa Luxemburg y Transnational Institute, 2013.

Andreucci Diego and Isabella Radhuber. "Limits to 'counter-neoliberal' Reform: Mining Expansion and the Marginalisation of Post-extractivist Forces in Evo Morales's Bolivia." *Geoforum*, Forthcoming.

Ballard, Richard, Adam Habib, Imraan Valodia, and Elke Zuern. "Globalization, Marginalization and Contemporary Social Movements in South Africa." *African Affairs* 104, no. 417 (2005): 615–634.

Bebbington, Anthony. "Underground Political Ecologies: The Second Annual Lecture of the Cultural and Political Ecology Specialty Group of the Association of American Geographers." *Geoforum* 43, no. 6 (2012): 1152–1162.

Bebbington, Anthony, Denise Humphereys Bebbington, Jefrey Bury, Jeannet Lingan, Juan P. Munoz, and Martib Scurrah. "Mining and Social Movements: Struggles Over Livelihood and Rural Territorial Development in the Andes." *World Development* 36, no. 12 (2008): 2888–2905.

Bridge, Gavin. "Mapping the Bonanza: Geographies of Mining Investment in an Era of Neoliberal Reform." *The Professional Geographer* 56, no. 3 (2004): 406–421.

Campbell, Bonnie. "Distancing from an Investment-led Perspective and Revisiting Past Approaches, Dichotomies, Categories and Boundaries of the Artisanal and Small Scale Mining Sector." *Third Word Thematics* 1, no. 2 (2016): 165–183. doi:10.1080/23802014.2016.1226145.

CINEP & PPP. *Minería, conflictos sociales y violación a los derechos humanos en Colombia. Segundo Informe Especial del CINEP* [Mining, social conflicts and human rights violation in Colombia. Second special report by CINEP]. Bogota: CINEP Programa por la paz, 2012.

Duarte, Carlos. *Situacion minera en los territorios campesinos, indígenas y afrodescendientes en el Norte del departamento del Cauca* [The mining situation in the peasants, indigenous and afrodescendants' territories from North of the Cauca department]. Cali: Instituto de Estudios Interculturales, Universidad Javeriana de Cali, 2012. http://www.aconc.org/wp-content/uploads/2015/04/Duarte_Carlos_IEI_Javeriana_Audiencia-17_Abril_2015_-contexto-minero-norte-del-Cauca.pdf (checked on the June 20, 2016).

El Espectador. "Ejército destruye seis retroexcavadoras en el norte del Cauca." *El espectador*, sección judicial 29 de octubre de 2014. http://www.elespectador.com/noticias/judicial/ejercito-destruye-seis-retroexcavadoras-el-norte-del-ca-articulo-524938 (checked on the February 20, 2015).

Elmundo.com. "Afrodescendientes no acompañarán la mesa de negociaciones en La Habana [Afrodescendants will not join the negotiations in Habana]." *Elmundo.com*, June 20, 2016. http://www.elmundo.com/portal/pagina.general.impresion.php?idx=276940 (checked on the June 22, 2016).

El nuevo siglo. "Crece aporte de minería al PIB [The mining GDB increases]." *El nuevo Siglo*, 6 de Diciembre de 2012. http://www.elnuevosiglo.com.co/articulos/12-2012-crece-aporte-de-miner%C3%ADa-al-pib.html (revised on the February 20, 2013).

Fals, Orlando. "Investigating Reality in Order to Transform It: The Colombian Experience." *Dialectical Anthropology* 4 (1979): 33–55.

Gordon, Todd, and Jeffery Webber. "Imperialism and Resistance: Canadian Mining Companies in Latin America." *Third World Quarterly* 29, no. 1 (2008): 63–87.

Gudynas, Eduardo. "La ecología política del progresismo sudamericano: los límites del progreso y la renovación verde de la izquierda [Political ecology of the Southamerican' progresismo: The limits of progress and the renovation of the green left]." *Sin Permiso* 8 (2010): 147–167.

Gudynas, Eduardo. "Transitions to Post-extractivism: Directions, Options, Areas of Action." In *Beyond Development. Alternative Visions from Latin America*, edited by Miriam Lang and Dunia Mokrani. Amsterdam: Fundación Rosa Luxemburg y Transnational Institute, 2013.

Hale, Charles. "Activist Research v. Cultural Critique: Indigenous Land Rights and the Contradictions of Politically Engaged Anthropology." *Cultural Anthropology* 21, no. 1 (2006): 96–120.

Harding, Sandra. "Rethinking Standpoint Epistemology: What is 'strong objectivity'." In *Feminist Theory. A Philosophical Anthology*, edited by Ann E. Cudd and Robinson O. Anderson, 218–229. Oxford: Blackwell.

Harvey, David. *The New Imperialism*. Chapter 4: Accumulation by Dispossession. New York: Oxford University Press, 2003.

Jimeno, Myriam, Daniel Varela, and Angela Castillo. "Experiencias de violencia: etnografía y recomposición social en Colombia [Experiences of violence: Etnography and social recomposition in Colombia]." *Soc. e cult., Goiânia* 14, no. 2 (2011), 275–285.

Katz, Cindi. "Vagabond Capitalism and the Necessity of Social Reproduction." *Antipode* 33, no. 4 (2001): 709–728.

Martinez Alier, Joan. "Los conflictos ecológico-distributivos y los indicadores de sustentabilidad [The ecological-distributive conflicts and the indicators of sustainability]." *Rebelión, Ecología social*, (2005).

Marx, Karl. *Capital I: A Critique of Political Economy*. Chapter 27: The Expropriation of Population of Agricultural Land. Moscow: Progress, 1887.

Mignolo, Walter. "Philosophy and the Colonial Difference." In *Latin American Philosophy: Current, Issues, Debates*, edited by Eduardo Mendieta, 80–86. Bloomington: Indiana University Press, 2003.

Ministerio de Minas y Energía. "Censo minero departamental colombiano [Mining census by department in Colombia]." http://www.minminas.gov.co/minminas/downloads/UserFiles/File/Minas/CensoMinero/CensoMineroDptal3.pdf (checked on the March 10, 2015).

Nem Singh, Jewwellord T. "Reconstituting the Neostructuralist State: The Political Economy of Continuity and Change in Chilean Mining Policy." *Third World Quarterly* 31, no. 8 (2010): 1413–1433.

Observatorio de discrimacion racial. *La disputa por los recursos naturales en los territorios afrocolombianos. El caso de Buenos Aires y Suarez (Cauca) desde una perspectiva de derechos humanos* [The dispute over natural resources in the afrocolombian territories. The case of Buenos Aires and Suarez (Cauca) from a human rights' perspective]. Bogota: Colección Justicia Global No. 5 UniAndes, 2011.

Offen, Karl. "The Territorial Turn: Making Black Territories in Pacific Colombia." *Journal of Latin American Geography* 2, no. 1 (2003): 43–73.

Perez, Mario. "Conflictos ambientales en Colombia: inventario, caracterización y análisis [Environmental conflicts in Colombia: Inventory, characterization and analysis]." In *Minería en Colombia. Control público, memoria y justicia socio-ecológica, movimientos sociales y posconflicto*, Dir. by Luis J. Garay Salamanca. Bogotá: Contraloría General de la Republica, 2014.

Proclama del Cauca. "Incautadas máquinas retroexcavadoras para minería ilegal en el norte del Cauca [Backhoes for illegal mining sized in the north of Cauca]." *Proclamadelcauca*, 31 de diciembre de 2014. http://www.proclamadelcauca.com/2014/12/incautadas-maquinas-retroexcavadoras-para-mineria-ilegal-en-el-norte-del-cauca.html (revised on the February 20, 2015).

Restrepo, Eduardo. *Etnización de la negridad: la investión de las 'comunidades negras' como grupo étnico en Colombia* [Ethnization of blackness: The investion of "black communities" as an ethnic group in Colombia]. Popayán: Universidad del Cauca & Pontificia Universidad Javeriana, 2013.

Tenthoff, Moritz. "Siguiendo la ruta de la minería en el Cauca [Following the mining rout in the Cauca]." *La silla vacía*, 14 de agosto de 2014. http://lasillavacia.com/content/siguiendo-los-pasos-de-la-mineria-en-el-cauca-48380 (revised on the February 20, 2015).

Vanguardia.com. "Defensoría llamó la atención de las autoridades por tragedia en minas ilegales [Ombudsman called the attention of the authorities after the tragedy in the illegal mines]." *Vanguardia.*

com, 1 Mayo de 2014. http://www.vanguardia.com/actualidad/colombia/257945-defensoria-llamo-la-atencion-de-las-autoridades-por-tragedia-en-minas-ile (revised on the May 26, 2015).

Velez-Torres, Irene. "Reflections on a Participatory Documentary Process: Constructing Territorial Histories of Dispossession among Afro-descendant Youth in Colombia." *Area* 45, no. 3 (2013): 299–306.

Velez-Torres, Irene, Sandra Rativa, and Daniel Varela. "Cartografía social como metodología participativa y colaborativa de investigación en el territorio afrodescendiente de la cuenca alta del río Cauca [Social cartography as a participative and collaborative research methodology in the upper basin of the Cauca river]." *Cuadernos de Geografía – Revista Colombiana de Geografía*, 21, no. 2, 2012: 59–73.

Velez-Torres, Irene, and Guillermo Ruiz. "Extractivismo neoliberal minero y conflictos socio-ambientales en Perú y Colombia [Neoliberal mining extractivism and socio-environmental conflicts in Peru and Colombia]." *Ambiente y Sostenibilidad*, 5, (2015): 3–15.

Different faces of access control in a Congolese gold mine

Sara Geenen and Klara Claessens

ABSTRACT
Mukungwe is a gold mine in eastern DRC, where mineral resources are commonly labelled 'conflict minerals'. Although Mukungwe's history has been characterised by conflict, we do not analyse the conflict here as a mere example of a mineral resource conflict. This article offers a more contextual understanding, starting from a detailed ethnographic and historical account and inspired by the literature on access control. In this sense, violence is understood as but one of the mechanisms of access control. Based on ethnographic and archival research, this article offers insights into the narratives and practices of the conflict's key players.

Introduction

The link between mineral resources and conflicts has been extensively discussed in the literature, starting with Collier and Hoeffler's thesis that the abundant presence of natural resources increases the probability of civil war, yet ultimately resulting in a body of mixed evidence.[1] Many authors however accept that the greed for mineral resources does play a role in the persistence, intensity or duration of conflicts as it changes the opportunity structure for armed actors and makes the continuation of war economically beneficial.[2] This is especially the case for 'lootable' resources that are easily accessible, relatively simple to extract and control and yield large financial returns.[3] The case of the eastern Democratic Republic of Congo (DRC) appears to neatly fit this model. In the late 1990s, mineral exploitation and trade became integrated in a 'war economy'[4] in which rebel groups, foreign armies and the national army have been fighting over control of mining sites and trade routes. As a consequence, the minerals that are found in this region (gold, tin, tantalum, tungsten) are frequently labelled 'conflict minerals'. Since the security situation is not yet stable and human rights violations are still legio, the 'conflict minerals' label is used until today, and has in recent years provoked the implementation of numerous legal and technical initiatives aimed at halting the trade in these contentious minerals, such as the American Dodd-Frank law or traceability and certification initiatives.[5]

Academic researchers, including the first author of this article, have argued that these legal and technical approaches are not only facing a number of practical challenges related to feasibility, accessibility and capacity, but they are also based on a misreading of the situation on the ground.[6] First, the greed discourse establishes a narrow causal link between minerals and conflict, but (a) this neglects other structural conflict causes and triggers (political, identity-based, land-related, etc.); (b) revenues from minerals fund activities other than conflict as well; and vice versa, (c) armed groups also derive rents from non-mineral activities such as road blocks, agriculture and trade in other commodities. Second, the restricted focus on conflict minerals tends to generalise from selected cases and to neglect local contextual elements. Third, it does not have enough consideration for the historical context in which informalisation and militarisation emerged, nor for broader governance dynamics.

This article aims to contribute to this argument and extend it by focusing on the history of Mukungwe, a conflict-ridden gold mining site in South Kivu province. Yet instead of analysing the conflict in Mukungwe as a mere example of a mineral resource conflict, this article offers a more nuanced and contextual understanding, starting from a detailed ethnographically and historically inflicted account and inspired by the literature on access, exclusion and control. In this sense, conflict and violence are understood as but one of the mechanisms of access control. In other words, revenues from minerals are used to establish and maintain control, but not necessarily through violence alone. As mechanisms may consist of both actions and speech, we focus our analysis on the *narratives* of the conflict's key players as well as on their *practices* and the strategies they use to establish control over resources and people. We consider narratives as utterances and discourses with a strategic goal. Practices are defined along the same line, as actions with political and strategic implications.

Our research is based on ethnographic fieldwork (October 2011 and May 2012), follow-up email conversations and archival data (documents, letters, court cases). Unravelling local histories in unstable institutional (post)-conflict environments such as eastern DRC is a true challenge. During our research, we were confronted with conflicting accounts, public and hidden power plays and strategic agendas. Therefore, our article does not intend to factually reconstruct history. We rather want to analyse actors' strategies and logic in order to make an argument about access, control and exclusion. We aim to demonstrate how their narratives and practices not only frame the conflict, offering us a detailed understanding of access and exclusion mechanisms, but also contribute to the rearrangement of power and authority structures. Such an analysis allows going beyond the simplistic causal link between the mineral extraction and conflict. By focusing on these local yet regionally connected power struggles, we try to offer a counter-narrative to the dominant ahistorical and reductionist understanding of conflict over mineral resources. In the next section, we introduce our conceptual framework. After that, we analyse different faces of access control across Mukungwe's history, using a chronological approach.

Access, control and exclusion

We take the literature on access, control and exclusion as a starting point. If access is the ability to benefit from a resource, control is the ability to facilitate or exclude other people's access. According to Hall et al.[7], all instances of land access imply some kind of exclusion. By focusing on ability, the attention is drawn to a wide range of power relations and norms that can constrain or enable actors to benefit from resources. 'Power' is thus crucial in this literature and its interpretation goes back to a Marxist tradition in political economy that differentiates between those who possess the means of production (the ruling bourgeoisie in capitalism) and those who labour with other people's capital or means of production (the proletariat). This is translated into a distinction between 'access maintenance' (the ability to keep your own resource access open) and 'access control'. Yet in access theory this does not automatically manifests itself as a class difference, as one individual may hold 'bundles of powers' which involve both mechanisms for maintenance and control.[8] Such mechanisms have been discussed in earlier applications of access theory in political ecology.[9] In this article we focus on the mechanisms for access control that are used by some of the protagonists in the Mukungwe conflict. These protagonists include customary chiefs, landowners, state agents and military actors; in short, we call them 'local elites'. They are able to establish a control over lucrative resources and exclude other groups from accessing these, a process commonly known as elite capture.[10] In turn, the revenues from these resources are used to maintain control, but also to consolidate their power base and try to establish – through a legitimation of their power – authority.[11] Land or resource conflicts are thus not solely about land and resources, but also about power and authority.

The question remains as to *how exactly* elites establish control and exclude people, in other words, what mechanisms they use. Two broad and to a large extent overlapping categorisations have been proposed. Peluso and Lund[12] identify four mechanisms: privatisation, legalisation, territorialisation and violence. Hall et al.[13] talk about market, regulation, legitimisation and force. Privatisation or market refers to the commodification of natural resources and the privatisation of public spaces. It creates exclusion through setting up financial barriers, creating enclosures and incentivising people to lay individual claims on land. Those who do not dispose of sufficient financial capital are excluded. Legalisation, or regulation, is in many ways closely connected to the dynamics of privatisation, but is more 'subtle'. It does not directly exclude people without financial capital, but rather those without legal titles that are backed by state law. Often, this comes down to the poor, the subordinate, those who do not have the capacity to conform to the rule of law. This may be because of financial reasons, but also because of social, cultural or political reasons, or because they do not have the necessary relations. So the adherence to the idioms of law and formality in fact legalises the dispossession of the poor. Legitimisation is a related mechanism, but it refers to the *moral basis* on which exclusionary claims are made. This way, it does not refer to legality, but to legitimacy.

Territorialisation refers to social actions and claims that aim at controlling space as well as people in that space, combining different techniques and policies of classification, registration and mapping.[14] In this sense, such claims and actions intend to 'governmentalize', in the sense of Foucault,[15] meaning to control and discipline the subjects within that space through a very complex form of power which is not imposed top-down, but which filters through society in the form of disciplinary institutions, knowledge and discourses. Discourses of 'identity', 'belonging' and 'ethnicity' may be part of this. According to Geschiere and Jackson,[16] the term 'autochthony', which has frequently been used in Sub-Sahara Africa, literally implies 'an origin of the soil itself' and is 'by inference a direct claim to territory'. It often translates into a claim to particular benefits or positions and is used as an instrument to mobilise support. The last mechanism implies the use of violence, force, coercion, threat or 'bare power', usually without a moral basis and thus not 'legitimate'. In the literature on resource conflicts, the violence mechanism usually takes centre stage, yet our analysis will illustrate how it is closely connected to all the other mechanisms. We will indeed refer to all abovementioned mechanisms, but there is one critical point that is not sufficiently addressed in these categorisations, namely the relational mechanism. Reviewing the recent literature on land control, Peluso and Lund[17] find that 'today's landlords are more likely to be corporate or state actors rather than local elites'. Although we agree that the actors involved in land control are very diverse, we also think it is important to highlight the fact that there are coalitions, networks of actors who work together in an attempt to access and control land. In this sense, local elites often work together with state and corporate actors, as our case study will demonstrate. The elites build up constituencies through mobilising popular support and making alliances with political and military power holders at the local, national and regional levels. Such alliances are mostly facilitated by financial means; in other words, they make use of what is commonly called corruption. Such alliances are also strategic and thus also shift in view of the broader political economy.

Setting the stage: Congo's mining history

Before digging into the history of the Mukungwe gold mine, we need to highlight some aspects of this broader political economy.[18] Congo's mining history has been characterised by both industrial and artisanal mining. Having started in the colonial period, industrial mining was responsible for a large share of the state's budget until the late 1970s; and since a few years this has again been the case. In the 1980s however, artisanal mining began to spread as a result of a massive economic and financial crisis caused by Mobutu's neo-patrimonial policies, combined with fluctuating international commodity prices. In 1982 it was also legalised. The 'rush' for the artisanal mining sites still intensified during the 1998–2003 war, when industrial mining had come to a complete standstill and artisanal exploitation provided a viable livelihood and in some cases also relative security, to hundreds of thousands of young men, but also women and children. At the same time, artisanal mining and trade became integrated in a regional war economy, as mentioned in the introduction. After the end of the war in 2003, large-scale mining companies, encouraged by the 2002 Mining Code that stimulated foreign private investment, regained their interest in the DRC's mineral deposits, first in the copper and cobalt-rich province of Katanga, later particularly in gold in the eastern Kivu provinces. In the meantime, artisanal miners are working in hundreds of sites which may lay outside or inside attributed industrial concessions. For many years, these artisanal sites have been governed by their own rules, based on a flexible interpretation of the law and on illegal taxation practices by local elites. Customary chiefs occupy a particular position as they still cherish their role as managers of the land – their position in land management is characterised by legal ambiguity, but in practice customary arrangements persist. Consequently, they commonly levy taxes and contribution on artisanal production and trade in minerals.[19]

Mukungwe is the name of a village, situated in the locality of Mushinga, in Ngweshe chiefdom. These administrative divisions coincide more or less with customary boundaries: a locality is governed by a *chef de groupement*; a chiefdom by a king or *mwami*. The superior administrative layers are the territory (in this case Walungu) and the province (South Kivu), which are governed by elected representatives.[20] Near the village, a miners' camp was built. It basically was a collection of a few hundred huts, restaurants, hotels and shops without any provision of water, sanitation or electricity; and was frequently referred to as Maroc. The population here consisted in majority of young men coming from neighbouring localities and chiefdoms, or from cities such as Bukavu, South Kivu's provincial capital. At the time of our fieldwork, 2011–2012, approximately 200 gold mining shafts, of which only a few were operational, were dispersed over two main hills: Kalanga and Lukwera (commonly called Kalazi). Along the Nyantanda river, alluvial gold exploitation was taking place. The yearly production was figured to be between 50 and 100 kg, with an estimated number of 3000–5000 artisanal miners working underground and in rivers.[21]

Faces of access control in Mukungwe

1970s to 1980s

Access to land was traditionally governed by the *kalinzi* system, a customary contract that implied tight hierarchical relationships between the *mwami* as custodian of the land and his subjects.[22] Adult men could give the *mwami* a symbolic payment, the *kalinzi*, mostly in the form of cattle. By doing so, they became political subjects of the *mwami* and obtained non alienable inheritable user rights over a plot of land. Nonetheless, user rights could also be taken away in case the land was abandoned or abused.[23] This centralised top-down governance structure created subjects and rigid patron–client relationships based on the distribution of and the access to land.

In the 1940s, Chunu Tukuwa is said to have paid the *kalinzi* to the *mwamikazi* (wife of the deceased *mwami* of Ngweshe chiefdom).[24] By this act, he acquired user rights over Lukwera hill, which were again confirmed by his son and heir Chunu Bagayamukwe at the time when Constantin Kurhengamuzimu had become chief of Mushinga locality. In addition, Chunu Bagayamukwe reportedly acquired the user rights over Kalanga hill. In this latter hill, gold was discovered in the late 1970s. Both Chunu – the customary land owner – and Kurhengamuzimu – the customary chief – immediately placed their claims on the hill. These claims were based on contentious interpretations of customary law. A first issue on which customary law did not provide a clear answer, was the question whether the payment of the *kalinzi* also gives the customary owner the right to extract from the subsoil. Second, we have earlier referred to the practice that customary chiefs share in the profits of gold exploitation and trade by levying taxes, which exists in many mines. Based on this practice, Kurhengamuzimu required Chunu to give him a share in the output.

But already in the early 1980s, the protagonists started to invoke state law and rely on official institutions as well. In July 1980, the chief of the local mining division – the public service in charge of administering the mines – wrote that two individuals supported by Kurhengamuzimu were exploiting gold in Mukungwe. He considered this problematic since Chunu 'enjoys the plenitude of his customary rights'[25] on this land. This framing is important, it is not clear what it means to 'enjoy the plenitude of your rights'. One year later, Kurhengamuzimu applied for a gold exploitation permit, as artisanal mining had now been legalised by Mobutu. Chunu however appealed this request, saying that Kurhengamuzimu 'stripped him of his rights whereas he filled all the customary requirements, particularly the payment of the *kalinzi*'.[26] The administrative authorities in Walungu territory seemed to agree with him, stating that 'there is no doubt that the contested land is the exclusive and incontestable property of citizen Chunu who fulfilled all the customary formalities towards the chief'.[27] They even suggested that Chunu 'may freely benefit from his land and be protected from the arbitrariness of the chief'.[28]

The way in which these state agents interpret customary law is remarkable. In 1973, a new property law had been introduced in Zaire, declaring all land to be state land and allowing the state to grant concession titles. While the law also foresaw particular arrangements for the management of land held under customary tenure, these never materialised and ambiguity persisted as to who could govern and use this land.[29] This resulted in the de facto persistence of the duality between state and customary land and the emergence of an amalgam of local hybrid arrangements. Vlassenroot and Huggins[30] described how this challenged the traditional power base of customary chiefs, yet provided them with new opportunities as well. Traditional patron–client relationships gradually eroded and were partly replaced by new forms of patrimonial relations based on economic gain and wealth accumulation. In this process, the meaning of *kalinzi* shifted 'from a perpetual symbolic contract between chief and peasant towards an individualized one-to-one purchase contract that can take place between any two parties'.[31] In this case, the state agents explicitly used the idioms of customary law; they interpreted Chunu's customary right as a private, exclusive title, while also condemning the 'arbitrary practices' of the chiefs.

In the late 1980s, the Mukungwe case was brought to court. After the high court's decision, which judged in favour of Kurhengamuzimu, Chunu lodged an appeal at the court of appeal in Bukavu in 1989. During the trial, Kurhengamuzimu denied that Chunu had ever acquired customary titles on the contested hills, while he did recognise him as land owner of a third hill.[32] According to the court of appeal's judgement, however, Chunu was 'the legal occupant of the contested land'. Kurhengamuzimu and his sympathisers were required to leave and pay compensations.[33] In a response to this judgement, Kurhengamuzimu wrote that the *kalinzi*, even if it were paid for Lukwera and Kalanga, had no value since these hills now fell under the mining legislation. Anticipating that, Kurhengamuzimu had applied for the above-mentioned exploitation permit, which he invoked again at this occasion. He thus tried to cover himself twice, saying that Chunu did not adhere to his customary obligations and if he did, that his official mining title now outweighed the concessionary title on the land. On the ground however, Chunu and his supporters continued mining. The group he gathered around him was called *Matonge*. Kurhengamuzimu blamed them for using the profits of the gold exploitation to corrupt the court of appeal: 'Chunu and his *Matonge* group are singing in the entire locality: with 20 *tolas* of gold we will win the case!'[34]

In this first part of the local history, all control mechanisms can be identified. It also becomes clear that they constantly overlap and are used in complementary ways. Legal pluralist theory already drew our attention to the co-existence of multiple sources of 'law' and authority.[35] These sources of law are ambiguous in themselves; they allow for different interpretations and may be used for strategic purposes. In the Mukungwe case, customary law did not provide a clear solution, but all parties made use of the idiom of customary law, so as to give their claims a 'legal' basis. In the case of the chief, there was a moral dimension when he referred to the fact that his subjects were obliged to pay rent. In other words, he considered this to be his legitimate share and he used it to establish authority over his subjects. In the case of the state actors, they talked about 'customary rights' as well, but interpreted these as private titles. This provided a good illustration of system-internal legal pluralism, where there are different, overlapping regulations with respect to land access within one legal field.[36] Yet claims were not only backed by state or customary law or by morality, but also by corruption, which we have conceptualised before as making use of financial resources to forge strategic relations. The '20 tolas' quote and the formation of this Matonge group indeed hinted at the importance of relational mechanisms, namely the formation of constituencies, groups of supporters or clients, as well as the use of mineral revenues to corrupt officials. As will become clear, all protagonists in the conflict have assembled such groups, who need to safeguard their interests and get protection in return.[37] Apart from corruption and territorialisation (the making of group identities), these groups also make use of violence, as one of their strategies.

1990s to 2003

In the early 1990s, the Kurhengamuzimu family retook the effective control over the mine.[38] According to the Chunu family, this was accompanied by excessive violence:

> Pasteur Kurhengamuzimu [son of the then chief of Mushinga] and his men have set 32 houses on fire, they looted 15 cows, 23 goats and 9 sheep. They took clothes, kitchen tools and other objects worth 2.851.250 zaïres at the time.[39]

On the other hand, there were indications that Chunu's support group Matonge was turning into an armed self-defence group. Acts of violence thus occurred on both sides. Meanwhile Chunu also tried to legalise his access to the mine, with success. A ministerial decree issued on 20 July 1993 granted him a two-year authorisation to prospect for gold.[40]

In the second half of the 1990s, local political struggles between both families linked up with the broader war context. In 1997, Laurent Kabila and his AFDL (*Alliance des Forces Démocratiques pour le Libération du Congo/Zaïre*) forces defeated Mobutu and seized power in Kinshasa. Yet Kabila soon turned against his former allies, asking his Rwandan advisors to leave the country and inciting hatred against the Rwandophones. Rwanda and Uganda then engineered a new rebel movement: RCD (*Rassemblement Congolais pour la Démocratie*), which occupied the eastern part of the country from August 1998 onwards. In reaction to this, local militia groups surfaced all over South Kivu. These militias were commonly called *Mayi-Mayi*, but had (and still have) very different characteristics. Whereas initially (before 1996) *Mayi-Mayi* had been set up as rural-based movements of self-defence against social and economic marginalisation, after August 1998 they started to make more and more strategic alliances with local, national and regional actors, often engaging in profit-seeking activities through the use of violence. The 'politics of belonging' were thereby used as an instrument to mobilise support among the population. Yet as Geschiere and Jackson[41] claimed, 'such manipulation can be successful only because it strikes such a deep emotional chord with the general population'. It is in this context that the emergence of the Bashi militia *Mudundu 40* or M-40 in Mushinga must be situated.[42] In 1999, M-40 emerged out of the existing private militia Kurhengamuzimu used to protect his claims to the mine. Whereas its official goal was to fight against the presence of Rwandan troops in South Kivu, it also aimed to keep control over the mines and to consolidate customary power. After an internal conflict over leadership mid-1999, the young Odilon Kurhengamuzimu[43] took the command. The urban Bashi elite – which was closely associated with civil society in Bukavu – supported M-40, considering it to be a useful instrument to politically position itself and to oppose RCD. During the years thereafter M-40 terrorised the entire region and kept its opponents in a notorious underground prison in Mushinga. In 2001 however, they allied with RCD because the latter had placed a Bashi governor in power. However, their turning sides and the atrocities they committed deprived them of the initial popular support they had, until their dissolution in 2003. During this whole period, the Chunu family resided in Goma in North Kivu,[44] where they managed to establish new alliances, as will become clear.

At the reunification of the country in 2003, some M-40 combatants were demobilised and integrated into the Congolese army, the FARDC. Other former rebels and soldiers started to work as artisanal miners. Mukungwe was placed under the command of the 10th military region, the FARDC regiment based in Bukavu. But these soldiers soon came to be actively involved in mineral extraction, with the complicity of Pasteur Kurhengamuzimu, as the bailiff of the court of appeal said.[45] The latter also concluded that the entire mining site had turned into a military camp and estimated that among the miners '90% are military elements, demobilized soldiers or deserters, and escaped prisoners'.[46] This statement may have been a bit exaggerated and only partly valid in 2011–2012. True, many artisanal miners in Maroc were former soldiers, but we also found many young men who could not find a job in Bukavu, or who were forced to abandon their studies for financial reasons. We found partisans of Chunu and Kurhengamuzimu, respectively. Others, mostly from outside Mushinga, were more neutral and sided with one party or another, depending upon their relative interests. As one local gold buyer told us: 'There is a hierarchy you need to see first. It all depends on who you find there [who is in control]'.[47] These miners were very mobile and moved from site to site in search for a high production.[48] Young men typically lived here alone, while their family was based in Bukavu or in one of the surrounding villages. They enjoyed the company of women who temporarily settled in Maroc, usually combining petty trade activities or restaurant or bar keeping with prostitution. Women from neighbouring villages walked the steep path over the hill every day, in order to sell food or drinks to the miners.

In this period the mechanism of violence became more prominent. Indeed, the conflict in Mukungwe became part of wider war dynamics, stretching out to the provincial and regional level. Seeing it the other way around, the story about M-40 shows how the 'Congo war', often called the 'first African world war', unfolded as a patchwork of localised conflicts with an extremely complex history.[49] This section also illustrated the importance of relational mechanisms. The conflicting parties each had their own constituencies, which were built on personal relationships and feelings of belonging. M-40, for example, explicitly presented itself as a Bashi self-defence militia fighting against foreign occupation. Its leaders derived their power from the mining rents and their authority from the discourse on 'defending autochthone interests'. Initially, this helped to mobilise popular support, but as we said, support diminished as a result of extreme violence and strategic political games. In Maroc there was also a group that was commonly called the 'diaspora' and was more neutral, shifting its position along with the power holder of the day. From their point of view, this was a strategy for maintaining access.

Towards the end of the described time period, the issue of 'conflict minerals' in Eastern DRC came to the fore. UN and NGO reports started to document how both government and rebel armed groups used mining rents to finance their war efforts, which fit in well with the greed theories we discussed in the introduction. Indeed, mining rents have been used to sustain conflict in the case of Mukungwe, and there has been frequent fighting over the mine. But this is only part of the story as revenues from gold mining and trade have also gone to support mechanisms of control other than conflict. Besides, placing the Mukungwe case in a broader historical perspective allows us to see that control mechanisms have not necessarily been violent.

2003 to 2010

The 2002 Mining Code[50] and revised mining policy created an institutional and legal framework that favoured industrial companies and promoted foreign direct investment.[51] For the government this was a way to better control the mining rents, although in practice negotiations over mining contracts often resulted in deals that were favourable for the companies and profits have not trickled down to the communities. In the artisanal mining sector, the government aimed at establishing a similar control and therefore a formalisation process had to be promoted.[52] According to the Mining Code 'artisanal exploitation zones' (AEZ) can be demarcated in areas where 'the technological and economic factors are not suited for the site to be industrially exploited'.[53] Miners can apply for a permit with one year validity to work in these zones, but they need to comply with the regulations on security, hygiene, water use and environmental protection specified in the 'code of conduct for the artisanal miner'.[54] Cooperatives can also obtain research permits within AEZ and should then ideally evolve into small-scale, semi-industrial operations. More recently all miners have been required to become member of a cooperative.[55]

In November 2005, Kurhengamuzimu obtained such a permit from the Provincial Mining Division, which is quite strange, knowing that there is no AEZ in Mukungwe. Apparently the permit was granted on the basis of his previous exploitation permit from 1981,[56] but this was disputed by Chunu and eventually cancelled by the provincial governor.[57] Kurhengamuzimu then decided to found a company, Samiki (*Société Agricole et Minière du Kivu*), and obtained two research permits from the national mining ministry.

However, the end of the war and the new investment climate also stirred up the interest of transnational mining companies. Both Chunu and his associates and Kurhengamuzimu and his allies tried to adapt to this changing reality by establishing relationships with interested transnational companies, while at the same time maintaining their strategies of violence, intimidation and militarisation at the local level. The most important new actor was the Canada-based multinational company Banro Corporation. The company claimed that Mukungwe was part of its Twangiza concession,[58] which was contested by most miners at the time of our fieldwork.[59] State law did not bring a solution at first, since the geographical coordinates in the Mining Registry were not clear about the exact limits of the concession. Banro allegedly concluded an arrangement with Samiki (Kurhengamuzimu) so that the multinational would take over the research permits, and Samiki could benefit by working as a subcontractor.[60] In 2008, Banro started building an exploration camp on Mwananzi hill, facing the hill of Kalanga. A similar process had unfolded in the neighbouring chiefdom Luhwindja, where Banro had started its exploration works in 2005.[61] This experience led the miners to fear that they would be chased from their shafts without appropriate compensation; and so they violently resisted Banro's arrival and set fire to the exploration camp. Banro at the time retreated together with the Kurhengamuzimu family. Chunu again took control, installing his military allies and delegating the daily management to two of their allies, Alexis Rubango and 'Marocain'. While it is difficult to find out how exactly the latter are related, both are also members of the local elite.[62] At this point in time they allied with Chunu, but as we will show later, they also strategically and opportunistically shifted their positions.[63]

The year 2008 was again characterised by violent confrontations between the different parties, both supported by armed groups.[64] These clashes had a devastating impact: at least one dead and one wounded person, burnt houses, pillages, lootings and expropriation of mining shafts. The behaviour of the Congolese soldiers was criticised by the provincial minister of justice and human rights. In a letter to the commander of the 10th military region, he explicitly asked for the replacement of the FARDC soldiers:

> I have the noble duty to ask you to replace the military currently quartered in Mukungwe by more disciplined and impartial troops; and to enforce a temporary suspension of mineral exploitation, until the quarrelling parties have acquired the necessary documents from the competent authorities.[65]

In his response, the commander of the 10th region argued that the presence of FARDC soldiers was absolutely necessary for the protection of the citizens and their belongings, especially in light of the nearby presence of the FDLR (*Forces Démocratiques pour la Libération du Rwanda*),[66] the Hutu rebels.

Meanwhile the Chunu family also pursued its legalisation strategies. In November 2008, they had several land titles registered in the official land registry.[67] At about the same time they tried to set up an industrial company Somimu (*Société Minière de Mukungwe*), after the example of Samiki, but that project failed. Chunu then founded a cooperative, the *Coopérative Agro-Pastorale et d'Exploitation Minière de Kalanga et Lukwera*[68] on 20 January 2009. In the course of 2009 attempts at rapprochement between both families were taking place. The Chunu and Kurhengamuzimu families were negotiating with Banro, which again provoked the resistance of many miners. The miners chased away Chunu and Kurhengamuzimu and remained on the side of Marocain, who disapproved of the agreement.

Kurhengamuzimu for his part, having secured the exploration rights for Samiki, was looking for an additional investor. Through a Bukavu-based broker with connections in Japan, he came in contact with Eco-Project, a Japanese group willing to invest in infrastructure, cement production, jatropha plantations and gold mining in South Kivu.[69] The position of *mwami* Ndatabaye of Ngweshe was used to convince Eco-project to invest, and after a visit of the *mwami* to Japan, the deal was concluded. Eco-Project, in which both Kurhengamuzimu and the *mwami* were associated as shareholders, acquired the research permits for four concessions bordering the Banro concession.

In the third part of this local history, we introduced companies and cooperatives. This can be understood in the context of the Congolese mining policy, which encourages industrial mining companies to apply for concession titles and artisanal miners to group in cooperatives. This created opportunities for local elites with good access to information and to financial capital, usually through their relations. They knew on which doors to knock in order to obtain a title and they established relations with (foreign) partners bringing in the capital. In other words, they used privatisation and legalisation mechanisms, which were facilitated through relations. We also found a narrative of territorialisation in the quote in which the military commander called to reinstall 'order', 'discipline' and 'competent state authority'. This again shows how revenues from mining have been used to establish control in many different ways.

2010 to 2015

But on the ground the situation remained unstable. Chunu built connections with Amani Leo soldiers based in Goma,[70] while Kurhengamuzimu got support from FARDC's 10th military region. Both groups alternately attacked Mukungwe, using a lot of violence and setting huts on fire.[71] In July 2011, Kurhengamuzimu assumed control with FARDC backing and installed a management structure locally known as the 'committee of Samiki', to which all miners had to pay contributions. But in October, Chunu launched another attack. This resulted in a new agreement between Chunu, Kurhengamuzimu, Rubango and *mwami*

Ndatabaye, who all agreed to share the rents collected from miners and traders. According to our interviewees, Chunu (together with Marocain) took 40%, Rubango 25%, Kurhengamuzimu 15% and the military 10%. Although this still reflected unequal power relations, the agreement did seem to create some calm. An interviewed miner said: 'Now they are working together, so we are also relieved. We are working towards the same objective'.[72]

Chunu and Marocain tried to organise and control the mining activities through the cooperative they had officially created in 2009, the *Coopérative Agro-Pastoral et d'Exploitation Minière de Kalanga et Lukwera*. The cooperative gave some technical support, but in return required all miners to become a member and contribute 10% of their production.[73] This provoked resistance from the part of a rival cooperative, as well as from a structure which called itself the 'committee' of artisanal miners.[74] One of the leaders explained why they created this committee:

> We established the committee because of the problems we had with the organisation of the site. The problem was that the *chef de colline* [Chunu] had his staff, who was there to protect his interests, so we decided to organize ourselves in view of defending our interests and harmonising our work. [...] The *chef* demanded that we pay the military, but how can we support soldiers; that is a task of the government. We understood that he wanted to do away with the committee, but we stood firm and said that we were not going to dissolve it.[75]

From the side of the government, there were several attempts to intervene and re-establish control, both through actions and through discourses. Although the place is very remote and not accessible by road, 'the state' was definitely not absent. State agents and soldiers were present to levy taxes. In 2012 the mining service Saesscam, whose task it is to support artisanal miners, was installed. Its representative was making an attempt to register all pits and miners, and levied (illegal) taxes as well. Yet in metaphorical terms, state actors have seen a need to 'bring the state back into' Maroc. The FARDC commanders, for example, repeatedly stated that 'state order' was to be reinstated. They have also pointed to the importance to remain in place for security reasons, even though according to the law a mining site should be secured by the mining police and not by FARDC soldiers. The same idea was expressed in a letter written by the administrator of the territory of Walungu (July 2011) concerning the 'restoration of state authority' in Mukungwe.[76] Among a number of ordinances consolidating power in the hands of the *mwami*, he also promulgated a prohibition on the sale of alcohol and an obligation for every inhabitant to register and pay a tax. The following quote testifies to the territorialisation strategy that was used:

> We noticed that many things are going on in Mukungwe without the knowledge of the political and administrative, even of the customary authorities. Roadblocks are for example set up on the road between Mushinga and Mukungwe without our consent. In order to reinstall state authority in Mushinga, the *chefferie* will proceed to identify all inhabitants of the *groupement*, as well as the activities they exercise.[77]

In 2012, the provincial government sent a delegation on the ground to map the concession borders.[78] On this basis, a ministerial decree determined that Mukungwe fell within Banro's concession. This incited some, including Chunu, Kurenghamuzimu and the miners' committee, to give up their resistance and join a negotiation process that was initiated and facilitated by the Bukavu-based NGO *Observatoire Gouvernance et Paix* (OGP).[79] Part of the miners however stayed and remained loyal to Rubango, who still did not agree with Banro coming in. On 1 April 2013, Pasteur Pasteur Kurhengamuzimu was shot at the doorstep of his house in

Bukavu. Rumours immediately spread about the role of his opponents, more particularly the Rubango family, but the perpetrators have not yet been found.[80]

In 2014, violence on the ground continued and made more than 20 victims. The negotiation process was heavily affected by this violence, as well as by the apparent political will to enforce decisions, such as the decision to launch a military operation against Rubango and his allies. High-ranked politicians in Bukavu are reportedly personally benefiting from Rubango's control over Mukungwe's gold production, reason for which the decisions to 're-establish state control' have so far remained idle words.[81] In September 2015, the miners' committee organised a sit-in at the Ministry of Mines, claiming a 'peaceful return of the artisanal miners, a reestablishment of order by the Congolese Government authorities, and a smooth continuation of exploration by Banro, as well as a resuming of the negotiations with the company'.[82] But neither this action, nor the multilateral negotiation process led to the desired result as events took yet another unexpected turn in late 2015.[83] A coalition of the 33th military region, the mining police and the intelligence service compelled Chunu and Rubango to reach an agreement by putting them in jail for almost a month. They came out with an agreement to again share the revenues from Mukungwe's pits according to the following percentage: 35% for Chunu, 30 for Rubango, 20 for Kurhengamuzimu, 10 for the security services and 5% for the community. In early 2016 activities in Mukungwe resumed, albeit not without tensions and difficulties, especially for the miners whose activities had been interrupted for several years.

In this section, we found that all control mechanisms were being continued along the same lines: the conflicting parties created cooperatives, made alliances, used violence and tried to legitimise their actions. It is now clear that privatisation is being promoted, but it is not yet clear what this would mean in terms of exclusion or dispossession of the local population. Whereas previously most miners opposed the company's arrival, part of them – represented by the committee – have now joined a negotiation process. Further research is needed to see how the configuration of control mechanisms will evolve, and what impact this will have on the lives of the people of Maroc.

Conclusion

We analysed the ongoing conflict in Mukungwe through the eyes of the key players. Their narratives and practices provided good illustrations of how control over resources, territories and people is established and maintained using mechanisms of privatisation, legalisation, territorialisation and violence, but also using relations and alliances.

As the Congolese mining policy favours industrial mining, a large part of the surface area in South Kivu has been given in concession to private companies. Local businessmen, customary chiefs or elites with financial capacity to do so have seized this opportunity to create companies and acquire concession titles, which they sometimes only hold for speculative purposes. This was the case for Kurhengamuzimu who created Samiki with the main intention of securing his interests and establishing himself as a viable player in the mining sector. These elites also position themselves as intermediaries between 'the community' and private investors, benefiting from their perceived legitimacy and/or their bare power, as Sonwabile Mnwana (this special issue) also demonstrates for the case of South Africa. In doing so, they are flexible enough to instrumentalise the changing institutional environment in order to consolidate control. The elites also used the official legal framework to consolidate their claims. This has been demonstrated with respect to the acquisition of official land titles and research and exploitation permits, the creation of cooperatives and the use of the courts' judgements. But the law itself has also proven to be dynamic, not always straightforward and open to interpretation. This has been displayed by the confusion about the designation of concessions to one or another company, or to artisanal and industrial mining. The more recent attempts to 'bring the site back under state's control' and to discipline its inhabitants have been analysed as territorialisation strategies. We have also evidenced how local elites have sought to establish and reinforce their control by connecting to, or even setting up, armed groups, and thereby inscribing themselves in broader war dynamics. Through this and a range of other examples, we have evidenced the importance of relational mechanisms.

By focusing on access control, we have provided less insight in access maintenance mechanisms by artisanal miners and other community members, those who are not part of the local elite.[84] It is clear that their responses have been fragmented: allying themselves to certain power holders, refraining from doing so, resisting against a planned privatisation or against ongoing violence (as Irene Vélez-Torres discusses in this special issue), or deciding to engage with the corporate-state coalition and get the best of out the deal.

There are some lessons that are applicable beyond this particular case. First, our analysis focuses on the key players' narratives and practices related to the conflict over land and resources, and shows how these are part of a process of claim making about legitimacy, power and authority. This allows for a more dynamic analysis of how power and authority are constantly (re)negotiated and (re)arranged. Second, these very 'localized' struggles do not occur in an a-political power vacuum, nor are they geographically restricted. These local conflicts are thus, in all their specificity, utterances of the broader political economy of conflict. Finally, our analysis offers a counter-narrative to the dominant discourse that establishes a causal link between conflict dynamics in the DRC and the extraction of mineral resources. Although it is true that revenues from mining have been used to directly finance conflict, they have been instrumental in establishing other forms of control as well. This article demonstrates that violence is only one possible strategy to maintain and control access. An overemphasis on the presence of mineral resources in combination with the use of violence as an access mechanism ignores the complexity of mechanisms available and thus ignores the 'different faces' of each specific conflict.

Acknowledgements

The authors thank the CEGEMI (Centre d'Expertise en Gestion Minière) team at the Catholic University of Bukavu and VLIR/UOS (Flemish Interuniversity Council) and IOB (Institute of Development Policy and Management, University of Antwerp) for funding the fieldwork. The Mukungwe case has also been studied in: Geenen, S. and Claessens, K. « De l'autre côté de la colline. Le contrôle contesté de Mukungwe, Sud-Kivu », in: Reyntjens, F., Vandeginste, S., Verpoorten, M. (eds.) *L'Afrique des Grands Lacs. Annuaire 2011-2012*, Paris, L'Harmattan: 181–203, 2012.

Notes

1. Collier and Hoeffler, "On Economic Causes of Civil War"; Blattman and Miguel, "Civil War."
2. Ballentine and Sherman, "Political Economy Armed Conflict."
3. Le Billon, "The Political Ecology of War."
4. For an analysis of war economies in eastern DRC, see Vlassenroot and Raeymaekers, "Conflict and Social Transformation"; Jackson, "Making a Killing"; and Nest et al., "Economic Dimensions War Peace."
5. The conflict minerals issue is highly topical and overviews of such initiatives are numerous, see Verbruggen et al., *Guide to Current Mining*.
6. For a critical approach, see Geenen and Custers, "Tiraillements autour secteur minier"; Perks et al., "From Discourse to Practice"; Geenen, "A Dangerous Bet"; Seay, "What's Wrong with Dodd-Frank"; Autesserre, *Trouble with Congo* and Johnson, "No Kivu, No Conflict."
7. Hall et al., "Powers of Exclusion."
8. Ribot and Peluso, "Theory of Access."
9. This includes literature on land and forests mainly, but also on water, fisheries, coral reefs, protected areas and topics such as climate change, community based resource management and land grabbing. Readers may want to start with Ribot and Peluso's seminal article and search for the articles that cite them; Ribot and Peluso, "Theory of Access."
10. Platteau, "Monitoring Elite Capture"; and Peters, "Inequality and Social Conflict."
11. Lund and Boone, "Land Politics in Africa." Authority is conceptualised as 'an instance of power that seeks at least a minimum voluntary compliance and is thus sought to be legitimated in some way'. Lund, *Politics and Dynamics of Property*," 3.
12. Peluso and Lund, "New Frontiers of Land Control."
13. Hall et al., "Powers of Exclusion."
14. Sikor and Lund, "Access and Property," 14.
15. Peluso and Lund, "New Frontiers of Land Control," 673.
16. Geschiere and Jackson, "Autochtony and Crisis Citizenship," 1.
17. Peluso and Lund, "New Frontiers of Land Control," 669.
18. Geenen, *African Artisanal Mining.*
19. Geenen, "Who Seeks, Finds."
20. The legal and administrative status of these decentralised entities is surrounded by a great amount of uncertainty and ambiguity. The constitutions of 2006 and two organic laws of 2008 attributed more fiscal, administrative and financial autonomy to these formerly deconcentrated entities. However, these were never fully implemented because of the weak administrative capacity of the Congolese Government. In some instance, the provisions in these laws were even deliberately blocked by the central government who fears losing control over the local level. The result is a great ambiguity with regard to the status of public authority at local level and the validity of their actions. Englebert, "Incertitude, autonomie et parasitisme."
21. Communication by cooperative leader, 29 February 2012 and another cooperative leader, 31 May 2012. See also BICC, "A la recherché de l'or proper."
22. Van Acker, "Where did all the Land Go."
23. Sosne, "Colonial Peasantization."

24. Letter by Chunu Bagayamukwe to the Governor of South Kivu, 5 July 2008.
25. Letter by chief of the local mining division to chief of the provincial mining division, 4 July 1980.
26. PV by the *commissaire de zone* of Walungu, 21 October 1981.
27. Ibid.
28. Ibid.
29. Mugangu Matabaro, "Crise foncière."
30. Vlassenroot and Huggins, "Land, Migration, and Conflict."
31. Utshudi Ona and Ansoms, "Reconciling Custom," 32; and Mugangu Matabaro, "Crise foncière."
32. Letter by Kurhengamuzimu Bafundisire to the president of the Court of Appeal in Bukavu, 2 October 1989.
33. Judgment RCA1630, Court of Appeal in Bukavu, 28 April 1989.
34. One *tola* is equal to 11.66 grams of gold. Letter by Kurhengamuzimu Bafundisire to the president of the Court of Appeal in Bukavu, 2 October 1989.
35. Griffiths, "What is Legal Pluralism?"
36. von Benda-Beckmann, "Who's Afraid."
37. Raeymaekers, "Protection for Sale?"
38. Interview, Mukungwe, 24 October 2011.
39. Letter by Chunu Bagayamukwe to the Governor of South Kivu, 5 July 2008.
40. Ministerial decree on attribution of research permit n.2107, 20 July 1993.
41. Geschiere and Jackson, "Autochtony and Crisis Citizenship," 3.
42. Vlassenroot and Van Acker, "War as an Exit from Exclusion"; Namegabe, "Pouvoir traditionnel"; and Lubala Mugisho, "Emergence d'un phénomène."
43. Son of Constantin Kurhengamuzimu, chief of Mushinga, brother of Pasteur and nephew of *mwami* Ndatabaye of Ngweshe.
44. Interview, Mukungwe, 15 February 2012.
45. Letter by bailiff of Court of Appeal in Bukavu, 30 April 2006. See also letter by first substitute of the prosecutor, 22 December 2006; Letter by chief of the office of the prosecutor and substitute of the prosecutor to governor of South Kivu province, 19 January 2007; Report by Lieutenant-colonel Kyembwa Wakagela of a mission to Mukungwe, 22 September 2007.
46. Letter by bailiff of Court of Appeal in Bukavu to the prosecutor, 23 June 2006.
47. Interview with trader, Mukungwe, 1 June 2012.
48. Interviews with miners, Mukungwe, October 2011 and May 2012.
49. Autesserre, *Trouble with Congo*; and Reyntjens, *Great African War.*
50. Government of Democratic Republic of Congo, *Loi n° 007/2002 du 11 juillet 2002 portant Code Minier* and Government of Democratic Republic of Congo, *Décret n° 038/2003 du 26 mars 2003 portant Règlement Minier*.
51. Mazalto, "Réforme des legislation minières."
52. For a critical discussion, see Geenen, "A Dangerous Bet" and Geenen, *African Artisanal Mining.*
53. DRC, *Mining Regulations*, 2003, T. 4, Ch. 1, Art. 109. Yet the law does also provide for a possibility to close down these zones if 'the factors justifying its creation ceased to exist', or if a 'new deposit necessitating large-scale exploitation has been discovered'.
54. DRC, *Mining Regulations*, 2003, T. 4, Ch. 1, Art. 111 and 112.
55. De Haan and Geenen, *Mining Cooperatives*.
56. Letter by Ntabala Chunu to administrator of Walungu territory, 7 November 2005.
57. Letter by provincial governor to administrator of Walungu territory, 5 November 2005.
58. For more details see Geenen and Claessens, "Disputed Access"; and Geenen, *African Artisanal Mining*.
59. Interviews in Mukungwe, October 2011 and May 2012.
60. Personal communication, 21 April 2012.
61. Geenen and Claessens, "Disputed Access." and Geenen, "Dispossession, displacement and resistance".
62. Marocain, for example, claims that he started working with Chunu while being a 'member of civil society' during the war period. He also claim that his father was the one who discovered gold in Kalanga hill: interview, Mukungwe, 15 February 2012.

63. Interview with a member of Kurhengamuzimu family, Burhinyi, 26 October 2011 and personal communication by email, 4 April 2012.
64. In a letter from July 2008, the Chunu and Rubango families denounce that they are constantly being attacked by Pasteur Kurhengamuzimu, who mobilises former rebels from Mudundu-40 who still hold their weapons. See also Global Witness, "Faced with a Gun."
65. Letter by provincial minister of Justice and Human Rights to commander of 10th military region, 24 February 2008.
66. Letter by commander of 10th military region to provincial minister of Justice and Human Rights, 11 March 2008; Letter by commander of 10th military region to provincial governor, 15 April 2008.
67. Certificates in the name of the Chunu family, the Rubango family and the Mweze family, both allies of Chunu, 10 November 2008.
68. Declaration of the foundation of the cooperative, 20 January 2009.
69. Website Eco-Project: http://ecoproject-group.com/mining-e.html.
70. Amani Leo (peace today) was, after Kimia I and II (peace and quiet) and Umoja Wetu (our unity) the military operation deployed in the Kivu's to fight against remaining rebels, particularly FDLR. The Amani Leo forces consisted of FARDC and former rebels, such as ex-CNDP (*Congrès National pour la Défense du Peuple*). It was suspended in April 2012. These operations were supported by Rwanda and – logistically – by Monusco, the UN peacekeeping force. According to the Group of Experts of the United Nations (United Nations, "Final Report"), Chunu discussed the Mukungwe case with the ex-CNDP leader Bosco Ntaganda himself, see Stearns, "From CNDP to M23".
71. Decision n.5072/001/BUR-AT/TER-WGU/2011 on restoration of state authority, 8 July 2011.
72. Interview with miner, Mukungwe, 2 June 2012.
73. Interviews in Mukungwe, May 2012. The June 2011 data from the Provincial Mining Division show that the file of Chunu's cooperative had been introduced for verification; see Provincial Mining Division, *Relève des cooperatives minières enregistrées ayant sollicité l'agrément par le biais de l'administration des Mines du Sud-Kivu*, 3 June 2011.
74. Interviews, Mukungwe, June 2012.
75. Interview with committee, 9 June 2012.
76. Decision n.5072/001/BUR-AT/TER-WGU/2011 on restoration of state authority, 8 July 2011.
77. OGP, "Rapport de l'atelier."
78. A first mission was cancelled because Banro had financed it, which was met with a lot of criticism from civil society.
79. Interview, Bukavu, 12 September 2014.
80. Personal communication, 6 April 2013.
81. Personal communication, September 2015.
82. Comité des creuseurs d'or du site minier de Mukungwe à Bukavu, *Mémorandum*, 13 August 2015.
83. Personal communication, 28 January 2016.
84. In Geenen, *African Artisanal Mining*, this point has been better developed.

ORCID

Sara Geenen http://orcid.org/0000-0002-0933-3974

Bibiliography

Autesserre, Séverine. *The Trouble with the Congo. Local Violence and the Failure of International Peacebuilding*. Cambridge: Cambridge University Press, 2012.

Ballentine, Karen, and Jake Sherman. *The Political Economy of Armed Conflict. Beyond Greed and Grievance*. Boulder, CO: Lynne Rienner, 2003.

von Benda-Beckmann, Franz. "Who's Afraid of Legal Pluralism?" *The Journal of Legal Pluralism and Unofficial Law* 34 (2002): 37–82.

BICC. *A la recherché de l'or propre. L'exploitation artisanale de l'or au Pérou et en Rd Congo*. Bonn: Bonn International Center for Conversion, 2012.

Blattman, Christopher, and Edward Miguel. "Civil War." *Journal of Economic Literature* 48, no. 1 (2010): 3–57.

Collier, Paul, and Anke Hoeffler. "On Economic Causes of Civil War." *Oxford Economic Papers* 50, no. 4 (1998): 563–573.

De Haan, Jorden, and Geenen, Sara. *Mining Cooperatives in South Kivu: Saviour or Extortionist?* IOB Analysis and Policy Brief 14. Antwerp: University of Antwerp, 2015.

Englebert, Pierre. "Incertitude, autonomie et parasitisme: les entités territoriales décentralisées et l'Etat en République Démocratique du Congo." *Politique Africaine* 125 (2012): 169–188.

Geenen, Sara. "A Dangerous Bet. The Challenges of Formalizing Artisanal Mining in the Democratic Republic of Congo." *Resources Policy* 37, no. 3 (2012): 322–330.

Geenen, Sara. *African Artisanal Mining from the inside out. Access, Norms and Power in Congo's Gold Sector*. Abingdon: Routledge, 2015.

Geenen, Sara. "Dispossession, Displacement and Resistance: Artisanal Miners in a Gold Concession in South-Kivu, Democratic Republic of Congo." *Resources Policy* 40 (2014): 90–99.

Geenen, Sara. "'Who Seeks, Finds': How Artisanal Miners and Traders Benefit from Gold in the Eastern Democratic Republic of Congo." *European Journal of Development Research* 25, no. 2 (2013): 197–212.

Geenen, Sara, and Klara Claessens. "Disputed Access to the Gold Sites in Luhwindja, Eastern Democratic Republic of Congo." *The Journal of Modern African Studies* 51, no. 1 (2013): 85–108.

Geenen, Sara, and Raf Custers. "Tiraillements autour du secteur minier de l'Est de la RDC." In *L'Afrique des Grand Lacs. Annuaire 2009–2010*, edited by Stefaan Marysse, Filip Reyntjens, and Stef Vandeginste, 231–258.Paris: L'Harmattan, 2010.

Geschiere, Peter, and Stephen Jackson. "Autochthony and the Crisis of Citizenship: Democratization, Decentralization, and the Politics of Belonging." *African Studies Review* 49 (2006): 1–7.

Global Witness. *Faced with a Gun, What Can You Do? War and the Militarization of Mining in Eastern Congo*. London: Global Witness, 2009.

Griffiths, John. "What is Legal Pluralism?" *The Journal of Legal Pluralism and Unofficial Law* 18, no. 24 (1986): 1–55.

Hall, Derek, Philip Hirsch, and Tania Li Murray. *Powers of Exclusion: Land Dilemmas in Southeast Asia*. Singapore: NUS Press and University of Hawaii Press, 2011.

Jackson, Stephen. "Making a Killing: Criminality and Coping in the Kivu War Economy." *Review of African Political Economy* 29, nos. 93–94 (2003): 517–536.

Jackson, Stephen. "War Making: Uncertainty, Improvisation and Involution in the Kivu Provinces, 1997–2002." PhD diss., Princeton University, 2003.

Johnson, Dominic. *No Kivu, No Conflict? The Misguided Struggle against Conflict Minerals in the DRC*. Goma: Pole Institute, 2013.

Le Billon, Philippe. "The Political Ecology of War: Natural Resources and Armed Conflicts." *Political Geography* 20, no. 5 (2001): 561–584.

Lubala Mugisho, Emmanuel. "L'émergence d'un phénomène résistant au Sud-Kivu 1996–2000." In *L'Afrique des Grands Lacs. Annuaire 1999–2000*, edited by Filip Reyntjens and Stefaan Marysse, 189–224. Paris: L'Harmattan, 2000.

Lund, Christian. *Local Politics and the Dynamics of Property in Africa*. Cambridge: Cambridge University Press, 2008.

Lund, Christian, and Catherine Boone. "Introduction: Land Politics in Africa – Constituting Authority over Territory, Property and Persons." *Africa* 83, no. 1 (2013): 1–13.

Mazalto, Marie. "La réforme des législations minières et Afrique et le rôle des institutions financières internationales: la République Démocratique du Congo." In *L'Afrique des Grands Lacs. Annuaire 2004–2005*, edited by Stefaan Marysse and Filip Reyntjens, 7–31. Paris: L'Harmattan, 2005.

Mugangu Matabaro, Séverin. "La crise foncière à L'Est de la RDC." In *L'Afrique des Grands Lacs. Annuaire 2007–2008*, edited by Stefaan Marysse, Filip Reyntjens, and Stef Vandeginste, 385–413. Paris: L'Harmattan, 2008.

Namegabe, Paul-Robain. "Le pouvoir traditionnel au Sud-Kivu de 1998–2003: rôle et perspectives." In *L'Afrique des Grands Lacs. Annuaire 2004–2005*, edited by Filip Reyntjens and Stefaan Marysse, 209–234. Paris: L'Harmattan, 2005.

Nest, Michael, François Grignon, and Emizet F. Kisangani. *The Democratic Republic of Congo: Economic Dimensions of War and Peace*. Boulder, CO: Lynne Rienner, 2006.

OGP. *Rapport de l'atelier d'harmonisation des points de vues des parties en conflit, pour un paix durable à Mukungwe*. Atelier organisé à Madaka, groupement de Mushinga du 13 au 14 juillet 2012, 2012.

Peluso, Nancy Lee and Christian Lund. "New Frontiers of Land Control: Introduction." *Journal of Peasant Studies* 38, no. 4 (2011): 667–681.

Perks, Rachel, and Koen Vlassenroot. "From Discourse to Practice. A Sharper Perspective on the Relationship between Minerals and Violence in the DR Congo." In *The Complexity of Resource Governance in a Context of State Fragility: The Case of Eastern DRC*, edited by Jeroen Cuvelier, 64–69. London: International Alert and International Peace Information Service (IPIS), 2010.

Peters, Pauline E. "Inequality and Social Conflict over Land in Africa." *Journal of Agrarian Change* 4, no. 3 (2004): 269–314.

Platteau, Jean-Philipp. "Monitoring Elite Capture in Community-driven Development." *Development and Change* 35, no. 2 (2004): 223–246.

Raeymaekers, Timothy. "Protection for Sale? War and the Transformation of Regulation on the Congo-Ugandan Border." *Development and Change* 41, no. 4 (2010): 563–587.

Reyntjens, Filip. *The Great African War*. Cambridge: Cambridge University Press, 2009.

Ribot, Jesse C., and Nany Lee Peluso. "A Theory of Access." *Rural Sociology* 68, no. 2 (2003): 153–181.

Seay, Laura. *What's Wrong with the Dodd-Frank 1502? Conflict Minerals, Civilian Livelihoods, and the Unintended Consequences of Western Advocacy*. Working Paper 284. Washington, DC: Center for Global Development, 2012.

Sikor, Thomas and Christian Lund. "Access and Property: A Question of Power and Authority." *Development and Change* 40, no. 1 (2009): 1–22.

Sosne, Elinor. "Colonial Peasantization and Contemporary Underdevelopment: A View from a Kivu Village." In *Zaïre: The Political Economy of Underdevelopment*, edited by Guy Grand, 189–210. New York: Praeger, 1979.

Stearns, Jason. *From CNDP to M23. The Evolution of an Armed Movement in Eastern Congo*. Nairobi: The Rift Valley Institute/Usalama Project, 2012.

United Nations. *Final Report of the Group of Experts on the DRC (S/2011/738)*. New York: UN Security Council, 2011.

Utshudi Ona, Innocent, and An Ansoms. "Reconciling Custom, State and Local Livelihoods: Decentralized Land Management in South Kivu (DRC)." In *Natural Resources and Local Livelihoods in the Great Lakes Region of Africa. A Political Economy Perspective*, edited by Stefaan Marysse and An Ansoms, 26–48. London: Palgrave Macmillan, 2011.

Van Acker, Frank. "Where did all the Land Go? Enclosure & Social Struggle in Kivu (D.R.Congo)." *Review of African Political Economy* 32, no. 103 (2005): 79–98.

Verbruggen, Didier, Evie Francq, and Jeroen Cuvelier. *Guide to Current Mining Reform Initiatives in Eastern DRC*. Antwerp: International Peace Information Service (IPIS), 2011.

Vlassenroot, Koen, and Chris Huggins. "Land, Migration and Conflict Tin Eastern DRC." In *From the Ground up, Land Rights, Conflict and Peace on Sub-Saharan Africa*, edited by Chris Huggins and Jenny Clover, 115–194. Nairobi/Petoria: African centre for Technology Studies/Institute of Security Studies, 2005.

Vlassenroot, Koen, and Tim Raeymaekers. *Conflict and Social Transformation in Eastern Congo*. Ghent: Academia Press, 2004.

Vlassenroot, Koen and Frank Van Acker. "War as Exit from Exclusion? The Formation of Mayi-Mayi Militias in Eastern Congo." *Africa Focus* 17, nos. 1–2 (2001): 51–77.

Artisanal gold mining in Kejetia (Tongo, Northern Ghana): a three-dimensional perspective

Esther van de Camp

ABSTRACT
Using in-depth ethnographic analysis of Kejetia's artisanal gold mining community in Tongo (Northern Ghana), this article argues that governance is based on artisanal gold miners' three-dimensional orientation towards both 'above-ground' social–political relations and regulations (a geometrically two-dimensional orientation) and the 'subterranean' geological sphere (which literally adds depth and a third geometric dimension). A substance relation to gold also means that its properties and geological context strongly influence miners' micro governance, as do miners' cosmological perceptions of gold's relationship to the above-ground and subterranean spheres. Hence, gold acquisition is embedded in a web of geological, social and cosmological relations beyond actual excavation. Artisanal mining policy and research should recognise the impact of this three-dimensional orientation and substance approach.

Introduction

Artisanal mining is a subject of interest in a diverse range of scientific and societal fields. This article aims to contribute to a more holistic and in-depth understanding of artisanal mining. This could provide beneficial insights for governmental organisations, environmental organisations, large-scale mining companies and social scientists, to name a few. Artisanal mining is dealt with and perceived very differently by different actors.[1] For example, from an environmental organisation's perspective, (artisanal) mining activities are viewed as a source of (unregulated) deforestation, contamination and (chemical) pollution of rivers, soils and vegetation.[2] National governments, in their turn, often associate artisanal mining with the occupation of land that does not administratively belong to someone. Or they view artisanal mining as profitable activities on which no taxes are paid – artisanal mining is in many places informally structured and not incorporated in the legal framework of a government.[3] Large-scale mining companies may view artisanal miners as difficult to deal with since they are sometimes active on the same ground – artisanal miners may have older, unacknowledged, land rights for areas that are granted as concessions to large-scale mining companies by the government. Conflicts over minerals and land tenure typically combine with, among other things, tensions between political manifestations of the state and the use of (local) 'history' to justify claims, and an increased local sense of identity and autochthony.[4] Artisanal mining is a very diverse practice; for example, it can take place on a very small or very large scale. It can be a side activity in the dry season when there are no yields from agriculture activities.[5] Alternatively, artisanal miners can be people who happen to live at a place where minerals are found. In other situations, people are driven by poverty to participate in mining activities, when there are few alternative livelihood possibilities.[6] Artisanal miners are often not included in decision-making processes. In order to obtain better insights into artisanal mining and be able to make adequate mining policies, it is important to engage with and research artisanal miners.[7]

A three-dimensional approach to artisanal mining is used in this article to enhance understanding of artisanal mining. Three-dimensional is meant in a geometric way: the 'above-ground' (two-dimensional) social–political arena on the surface is studied, as well as the subterranean sphere which literally adds depth and a third dimension.

The approach in this article is inspired firstly by the concept of the *social–ecological system,* developed by Ostrom,[8] and secondly by anthropological literature insisting on researching the geological and substance influences on social organisation around artisanal mining sites.[9]

A social–ecological system is a dynamic complex adaptive system in which resource flows and uses (natural, socioeconomic and cultural) are regulated through a combination of social and ecological systems. In a social–ecological system, 'the natural context' of people involved in common pool resource-related activities should be researched as both an antecedent and consequence of social interaction.[10] To wit, ecosystems and social action are constantly influencing each other and they should not be researched separately, but as dynamic interactions within a social–ecological system.[11,12] Anthropological fieldwork could provide such research about society–environment interactions for common pool resource systems.[13] Anthropology's in-depth fieldwork methodology, long engagement in questions of society–environment interactions and broad, holistic view of society have already been shown to yield valuable insights into the science, impacts and policy of climate change.[14] These insights demonstrated the strong mutual influences between human action and ecosystem dynamics, and the need for a more thorough understanding of these connections between human actions and ecosystem dynamics in general.[15]

More specifically for this research, anthropological studies have begun to demonstrate the involvement of geological analysis in the study of social structures around artisanal mining. Artisanal mining activities are very mineral specific,[16] and they are highly dependent on the geological structures underground. In 1985, Godoy suggested integrating geological knowledge with economic and sociocultural knowledge of mining.[17] However, most anthropological and social research on mining stayed above ground. Gewald, in 2010, attempted to support interdisciplinary anthropological, historical and geographical perspectives on gold mining in West Africa by providing an analysis of gold as a geological item.[18] Panella also refers to the importance of a three-dimensional and interdisciplinary perspective on mining:

> Anthropological research provides empirical insights into social processes. Secondly, historical research allows for long-term perspectives with which to counter pre-suppositions of empty lands and institutional vacuum. Thirdly, scrutiny of specific sites where both artisanal and industrial mining takes place, shows not just how these ways of mining compete but are also complementary in the three dimensional spaces of co-habitation. Together, this interdisciplinary

knowledge allows investigating broader articulations of mining with other forms of land use as part of long term, dynamic processes of cohabitation.[19]

Luning argues that there is often a form of cohabitation on or around mining sites.[20] This cohabitation refers to the spatial intertwinement of activities around artisanal mining and large-scale mining, and/or with agricultural activities and settlements. In the quote above, Panella introduces the concept of a 'three-dimensional' approach to mining and points out that to understand the dynamics of cohabitation in mining, anthropological and historical research should be supplemented by geological research of the mining site concerned. Luning and Pijpers also emphasise that in cases of cohabitation and in the analysis of relations between multiple actors around gold mining, a three-dimensional approach is vital.[21] They show that mining companies can use geological knowledge, in combination with knowledge of artisanal gold miners' struggles with the water level, in company–community strategies. In their upcoming article, they make an interesting analysis of a mining company that used geological knowledge to decide how to govern the concession: artisanal miners 'are given' the upper layer of the ore body, and the mining company knows that it will be able to mine what is underneath the water table in the future. Luning and Pijpers relate this strategy to above-ground pressures in the social field and to subterranean features and the idea of 'vertical reciprocity'.

Different ecological and geological structures require different ways of mining that may have different impacts on the local social–ecological system dynamics. In nature, there are two types of gold: alluvial gold that is often found in river beds (dry gold with about 90–95% purity) and amalgam gold that is located in ore (about 82–90% purity).[22] According to Aryee, methods used in the artisanal and small-scale mining (ASM) sector of particularly gold in Ghana can be placed into three groups: shallow alluvial mining, deep alluvial mining and hard rock (lode) mining.[23] Shallow alluvial mining refers to the 'dig and wash' technique, and the hard rock (lode) mining refers to amalgam mining – the latter is discussed in this article. Artisanal miners and large-scale miners may have interests in the same land, but the geological characteristics of the gold in the ground that are suitable for their techniques may differ. Bleischwitz et al. point out that large-scale mining companies mainly need large amounts of minerals in the ground in order to make the settlement of industrial installations profitable,[24] while artisanal miners benefit more specifically from high concentrations rather than large amounts of minerals in the ground. Banchirigah and Hilson emphasise that in Sub-Saharan Africa, artisanal mining activities mainly involve substance stocks that are relatively close to the surface.[25]

This article focuses on the three-dimensional perspective discussed above on mining on a microlevel. Ballard and Banks identify three main kinds of stakeholders involved in mining: governmental agencies, mining companies and local communities.[26] All three were part of the research setting for this article.[27] However, this article does not assess the three-dimensional relations between these stakeholders. Instead, it takes a three-dimensional approach and zooms in on microgovernance in Kejetia's artisanal mining community. It argues that microgovernance in Kejetia's artisanal gold mining community is based on a three-dimensional orientation of artisanal gold miners and on their substance approach to gold. Orlove and Caton emphasise that a particular piece of matter can be seen by different actors in different ways. Therefore, they mention that matter is approachable as both a resource and a substance.[28] A resource view only refers to the economic value of matter, while a substance view allows for social, cultural, cosmological and geological perceptions as well. These perceptions can differ in time and space and among different actors, and they are worth researching because they influence the way in which actors orientate and organise themselves around the substance. Approaching gold as a substance can take account of both above-ground and subterranean elements that structure artisanal mining; cosmological perceptions of gold, just like the geological context and substance features of gold, are part of such a substance approach and of a three-dimensional orientation.

Microgovernance is in this article defined as the way in which social actors – individuals, teams or organisations – coordinate their activities (formally and informally) to deal with collective problems or to achieve a common goal. Coordination is the way in which social actors adjust their activities to one another.

This article is based on multi-sited ethnographic field research among artisanal gold miners in Kejetia and the political elite of the Tallensi in Tongo.[29] Below, the research case is first introduced, including the complexity of the local social–political relations that artisanal gold miners in Kejetia have to deal with due to the many different actors that have made claims on Kejetia's gold and land over the past two decades. Then, the microgovernance of artisanal gold miners in Kejetia is discussed through a description of the implications of starting and running a *ghetto* in Kejetia (ghetto is the local English term for a mine) and processing the gold ore. Added to this description are cases of artisanal miners' strategies, thus illustrating both the third dimensional (subterranean) orientation of artisanal gold miners and the details and nuances of above-ground spatial (two-dimensional) governance in Kejetia. The article concludes that policy-makers and researchers that deal with artisanal mining should recognise the impact of the three-dimensional orientation and substance approach of artisanal miners on their mining behaviour.

Kejetia: a local mining history

Kejetia is an artisanal mining area in the village Gbani in Tongo (the administrative capital of the Tallensi district) in north-eastern Ghana.[30] Kejetia and neighbouring mining communities were established in the mid-1990s.[31] Kejetia, though situated among sparsely populated farmland and bush, resembles a small town, and gold mining takes place among mud-block homes and a variety of small businesses, including food vendors, pito (millet beer) brewers, seamstresses, petty traders and mechanics.[32] An official with the Minerals Commission estimated that the 2010 population was about 2500, although the transient nature of mining communities makes this difficult to approximate.[33] Both men and women participate in above-ground artisanal mining activities and enterprises and above-ground gold processing activities in Kejetia. Typically, (young) men in Kejetia participate in artisanal mining activities to earn money to build a home or marry a wife. Also, both men and women typically attempt to earn money to go to school or send their children, brothers or sisters to school. Only men go under the ground to mine gold, which is explained by many people in Kejetia, both men and women, in terms of physical strength and fear. This is unlike explanations in literature about artisanal mining in Burkina Faso, where this is attributed to the fact that women's traces of sexual interaction or menstrual blood might chase away the gold.[34] Many miners state that their work and residence in Kejetia are temporary; however, these miners often have been working in Kejetia for many years, sometimes since the origin of Kejetia. Over the past two decades, many different actors have made claims on the gold and land in Kejetia. Artisanal gold miners in Kejetia therefore have to deal with complex local social–political relations.

In 1984, the Tallensi king and paramount chief of Tongo founded Gbani and *enskined* (made chief) the Gbani chief (who remains the Gbani chief to the time of writing).[35] Ten years later, in 1994, gold was found in Gbani and the gold mining community Kejetia emerged. Now, people from all over Ghana as well as from Burkina Faso, Togo, Benin and Nigeria are involved in artisanal mining activities in Kejetia. All of these people that are coming, staying and going are together part of Kejetia; however, many informal power relations and opportunities to acquire access to gold are also explained by artisanal gold miners in terms of who is autochthonous and who is 'a stranger' in Kejetia. Artisanal miners who are from Tongo and belong to the Tallensi are seen as autochthonous by both artisanal miners and the local authorities. Artisanal miners from other parts of Ghana and neighbouring countries are often perceived to be strangers. People from Tongo, and especially from the chief family from Tongo, have a powerful position in Kejetia due to their strong relation with the representative chief in Kejetia, who is a local authority in Kejetia – this is further explained below. When the artisanal gold mining activities in Kejetia came to be established, another powerful actor entered the picture.

Artisanal miners narrate that in 1996, the Minerals Commission 'discovered' the artisanal mining activities in Kejetia and proclaimed that artisanal miners had to apply for a small-scale mining concession (and that in 2013, artisanal miners were again obliged to register).[36] This happened following the policy to legalise the small-scale and artisanal mining sector.[37] Through a system of registration, Ghanaian prospective artisanal gold miners are required to apply for a license for entitlement to work a 25 ha plot of land.[38] This obligation relates only to Ghanaian miners, which points to the Minerals Commissions' broader understanding of all Ghanaians, regardless of their being autochthonous or stranger, as being equally entitled to make claims on the gold in Kejetia. Ghanaians who are originally strangers in Kejetia use in particular these concessions to legitimise their claims on the gold and to acquire a more secure position in Kejetia. The application for a small-scale mining concession needs to be addressed to national governmental agencies. To wit, in Ghana, the state is the ultimate owner of all underground mineral wealth.[39] However, artisanal gold miners often have to deal with land surface rights that are locally regulated based on long historical practices and land rights. This requires artisanal gold miners to orientate themselves within complex, multidimensional social–political regulations; subterranean minerals and land surface have different lines of authority. This is also the case in Kejetia and is further discussed below. The coming of the Minerals Commission gave rise to a new sort powerful actor in Kejetia: the *concession leader*. Concession leaders are artisanal miners who have received a small-scale gold mining concession from the Minerals Commission and sell or sublet pieces of their land to other artisanal miners in Kejetia. Most, but not all, concession leaders in Kejetia are from Tongo; foreigners often legitimise their claims on gold through concession leaders.

In 2008, the Shaanxi Mining Company (GH) Ltd. (a foreign, service-providing, Chinese mining company) started prospecting in Gbani and in 2011, the company started operating. Since 1983, many foreign mining companies have started operating in Ghana, as a result of the Ghanaian national Economic Recovery Program (ERP).[40,41] Shaanxi provides services on the concession of the son of the Gbani chief: a representative chief in Obuasi (a neighbouring gold mining community to Kejetia).[42] Due to a lack of communication (by the representative chief and the company), the coming of Shaanxi was accompanied with dissatisfaction, rumours and confusion among the artisanal miners in Kejetia, which is partly localised in their perception of subterranean geological structures.

Artisanal gold miners, not being involved in the decision-making process and agreements with Shaanxi, expected that Shaanxi would cooperate with them and bring developments to Kejetia such as economic growth, water supply and sanitation. These expectations were not met. When Shaanxi built a bridge from Kejetia to another gold mining community in 2013, artisanal gold miners perceived this as a threat: an attempt to create ways to mine outside their concession. The artisanal gold miners demonstrated, showing banners that expressed their dissatisfaction and blowing up one of Shaanxi's industrial machines. Since then, a police team visits Kejetia every week. Artisanal gold miners claim that the tension has decreased due to the police teams and the fact that the boundaries between Shaanxi and the artisanal mining (both underground and above ground) have become clearer. However, artisanal gold miners in Kejetia still very much perceive an unequal power relation between themselves and Shaanxi: they say that, '*the Chinese* have machines while we have only man power'. Some emphasise their expertise in (organising) mining activities and pursue a similar alignment to the one that the Obuasi representative chief made with Shaanxi: 'if I could have [private or state] investors and service providers for my concession, I would make the same deal'. Artisanal gold miners' (social–political) perceptions of their power in relation to Shaanxi are also embodied in their interpretations of subterranean geological structures. For example, some artisanal gold miners in Kejetia are convinced that Shaanxi built a concrete wall under the ground to block the way for artisanal gold miners and to create more mining space for the company.

The current Tallensi king and paramount chief of Tongo (since April 2015; at the time of the research, he was the supreme prince of Tongo) is also dissatisfied with Shaanxi. He is convinced that the land and gold in Tongo originally belong to the Tallensi and to his father, who founded Gbani. The Tongo chief believes that the arrangement with the foreign mining company is disadvantageous for 'his people'. The claims that he makes on Gbani's gold (mining communities) are in terms of autochthony arguments and the issue of 'who was here first'. Here, the paramount chief of Tongo struggles with the subsoil mineral ownership of the Ghanaian state while he makes political claims on Kejetia's land and subterranean gold based on long historical practices and land rights. This use of autochthony claims in terms of 'who was first' cannot be seen as classical or traditional, but it is a typical, modern, phenomenon of the twenty-first century as part of the implications of globalisation.[43] As more and more people are able to legally claim local resources (such as land or minerals), people (all over the world) tend to emphasise their autochthony and use this as to legitimise first rights over these resources. This is visible in Tongo, as a result of migration to the mining area and the privatised mining sector in Ghana. As described in the next paragraph, artisanal gold miners who are from Tongo also have more power in Kejetia than people who are not from Tongo. Artisanal miners who are not from Tongo, in their turn tend to legitimise their claims on land through the legal framework of the national government, even when stating that they do not view the government as a relevant, present or visible actor in Kejetia – many artisanal miners (both from Tongo and from elsewhere) perceive Kejetia as its own country.

Above-ground social–political relations in Kejetia have been complex and dynamic in the described local mining history of Kejetia. Next, the article shows how the three-dimensional (above-ground and subterranean) orientation of artisanal gold miners relates to microgovernance in Kejetia.

A three-dimensional orientation: starting a 'ghetto'

Artisanal gold miners in Kejetia base their mining strategies on a three-dimensional orientation. The complex combination of an underground geological orientation and an above-ground social–political orientation originates with the beginning of a *ghetto* (*ghetto* is the local English term for a mine).

A strategy for beginning a *ghetto* is geological orientation, prospecting for gold. Multiple rows of small holes in the ground (rows of ghettos) are visible on Kejetia's land. Artisanal gold miners in Kejetia know that gold is usually in the ground in lines (gold veins). They know that ghettos in one row probably lead to one subterranean gold vein. The best way to find a good place to mine is thus to start by exploring the ground at the ends of the ghetto rows, where there is a good chance that powder gold (small amounts of gold dust) can be found that will lead to a gold vein.

Above-ground orientation is combined with the geological orientation; artisanal miners need to socially and politically position themselves in order to facilitate their mining activities. Who can give them permission to use the targeted land or to extract gold from the ground? In Ghana, subterranean mineral ownership and surface land ownership are regulated through different actors and regulations. The state is the ultimate owner of subsoil mineral wealth.[44] However, artisanal miners often deal with 'surface rights' that are locally regulated and based on long historical practices and land rights.

To use land in Kejetia, artisanal miners need permission from Kejetia's representative chief and representative *tindana* (spiritual land owner) – from now on, they are called simply 'chief' and 'tindana' in this article.[45] In return for permission, artisanal miners will give them a portion of the profit. The relationship with the chief is valuable for artisanal gold miners. To wit, the chief is the one who mediates when conflicts arise between artisanal miners; these conflicts often involve subterranean encounters of artisanal miners. The tindana mediates in cases of geological or spiritual trouble with the underground (such as a collapsed ghetto or a lack of gold in the ground) by sacrificing a chicken, goat or cow to satisfy the spirits and ancestors of the land – below, this article elaborates on artisanal gold miners' cosmological perceptions of the relations between the tindana, gold, spirits and ancestors. Subterranean encounters, geological matters and spiritual matters are thus part of artisanal miners' above-ground social–political orientation with regard to land use.

To extract gold from the ground, artisanal miners also need a concession for small-scale mining from the Minerals Commission (as discussed above). However, only Ghanaians can apply for a concession[46]; foreigners have to find other ways to begin artisanal mining. In Kejetia, concession leaders play a large role in artisanal gold miners' access to subterranean gold. Concession leaders usually own about 25 acres (about 10 ha) of land. One way for foreigners (or other miners) to enable mining activities is to ask the concession leader for a part of the land in exchange for a share of the profits.[47] The complex multidimensional regulations, in which artisanal gold miners have to engage, make it all the more important for artisanal gold miners to position themselves well, socially and politically, towards powerful actors such as the chief, the tindana and concession leaders.

Issues of access to subterranean gold can highlight informal power relations between artisanal gold miners in Kejetia, relations that are, as mentioned above, for a large part explained in terms of people 'being from here' (being autochthonous, from Tongo) or 'being a stranger'. For example, artisanal mining teams from Tongo have a stronger voice with the mediating chief when they encounter with other mining teams under the ground – such underground encounters are caused by different mining teams following the same gold veins. Furthermore, Kejetia's committee of concession leaders mostly consists of concession leaders from Tongo. A concession leader from Bolgatanga remarks that some concession leaders from Tongo abuse their power in subterranean competition or conflict.[48] Therefore, this concession leader deliberately joined the committee of concession leaders to strengthen the power of people from outside of Tongo. A distinction between autochthones and strangers seems to be emphasised in artisanal miners' stories of subterranean competition.

Artisanal miners' intentions to comply with above-ground social–political regulations, mainly with the legal framework of artisanal mining, are sometimes challenged by their three-dimensional orientation. Artisanal miners state that they want to mine within the legal framework and that many mines in Kejetia are legal.[49] However, the process of getting permission for a concession is complex and takes a long time and artisanal miners cannot always ignore geologically urgent situations (especially when poverty is pressing); during above-ground social–political arrangements, a subterranean geological issue can arise. To illustrate this, an artisanal miner might be waiting for permission for a concession while the mining team of the ghetto right next to his land is suddenly finding lots of gold. Then, there is a good chance that the other mining team will soon come close to the ground under his land by following the gold vein. The artisanal miner cannot risk other miners getting to the gold under his land: he has already invested in relations with the chief and tindana and in exploring the ground of that land, so he cannot afford to wait and lose what is possibly 'his' gold. In cases that ghettos from different mining teams meet under the ground, the mining teams mostly come to an agreement or ask the chief to intervene. However, if a miner is not yet under the ground, there is a good chance that the other mining team will just keep following the gold vein under his land. The geological orientation of artisanal gold miners in Kejetia can, in these cases, overshadow their social–political orientation.

Going under the ground

Subterranean artisanal mining activities in Kejetia are a result of miners' three-dimensional orientation and their substance approach to gold. This is argued and illustrated with an explanation of microgovernance of underground mining activities.

When an artisanal gold miner in Kejetia has permission to mine (has made an agreement with the chief, tindana and the Minerals Commission or a concession leader), he needs a team for his mining activities. All artisanal miners in Kejetia work under the ground in teams – although many artisanal gold miners in Kejetia perceive mining as an individual 'everyone on their own' process: 'if you are not there, you will not get your share'. These teams for subterranean activities only include men and can consist of 10–100 miners. The organisation of subterranean mining tasks and teams depends on the subterranean geological structures that are found. The general tasks in a mining team are: sponsor, ghetto owner, gang leader,

Table 1. Names and descriptions of geological features by artisanal gold miners in Kejetia.

Name	Description
Stone	Stones that contain gold; ore
Rock or waste	Stones that do not contain gold
Gold-rock	Layered stones with both gold-containing layers and 'no gold' layers
Bulk	Huge piece of gold-containing stone (*blast man* needed)
Package	Many small pieces of gold-containing stone (*chisel man* needed)
Gutter	Gold line, both vertical and horizontal
Rainbow	Bent gold line
Table	Square of gold-containing rock (which you stand on)
Lane	Horizontal *ghetto* following a horizontal gold line (like a subterranean street)
Ceiling	Stone that contains gold (vein) (above your head)
Powder gold or powder money	Gold dust in ore (small amounts)
Nugget	Big pieces of gold in ore

blast man, drill man (*moyee man*), chisel man (*chiseler*), security man and local boy. Not every team has the same structure, but most teams are very similar.

The *owner* of the concession can also be the owner of a ghetto. A ghetto owner has to take care of his mining team; arrange the mining tools; and give his workers money for food. This is an expensive responsibility, especially in the beginning when the big nuggets of gold have not yet been reached and there is only some powder gold in the stones (Table 1). In such a situation, ghetto owners often decide to take a *sponsor*. The sponsor will invest in the mining activities and over the years, the ghetto owner will pay him or her back, with interest. Furthermore, he has to sell his gold to his sponsor (who is also his *buyer* now) for a relatively small price. The ghetto owner might choose to also take up the role of *gang leader*, which means that he is the one who gathers the required workers when certain activities need to be done. This can be a complicated responsibility because many miners move frequently between Kejetia and their home towns; not all team members are always in Kejetia because their family might be in their home towns or villages.

The first hole in the ground is made by the *drill man*. He drills a hole in the surface in the shape of dynamite. Then, the *blast man* uses dynamite, electric wire and a wick to blast away the first stones. With this method and by following the powder gold, a square or round vertical hole (roughly one and a half meters in diameter) is shaped: a ghetto.

Under the ground, a mining team will find different kinds of stone. Artisanal gold miners in Kejetia have a rich terminology for subterranean geological features. They define subterranean geological structures by their containment of gold, the direction of gold veins and by the position of gold-containing stones in relation to the position of the artisanal gold miners themselves (Table 1). Artisanal gold miners in Kejetia base their day-to-day mining strategies on their understanding of the subterranean geological structures and cosmology.

During the digging process, an artisanal miner has to keep orientating himself three-dimensionally; he has to anticipate and react to social above-ground factors, geological subterranean factors and also to cosmological relations. An artisanal gold miner illustrates this with the following experience:

> Currently, our team has a problem. Our ghetto owner has always been both our ghetto owner and our sponsor. He had a lot of money, but it vanished. He said that he had to help some people, maybe his family. But now, the people don't want to help him back. Our ghetto owner doesn't want to take a sponsor, because he has always done it alone and he doesn't want to be dependent. So there is no money to eat or work, but the work is there – there is gold in the

> ground! I called all the local boys together to discuss. We decided to all disappear for a while, to go on a strike. But now the situation is pressing, because we realized that we have reached a Lane (Table 1)! We know that the Lane has a lot of gold, because we saw other mining teams that reached that same Lane taking a lot of money. We have to hurry, we don't want other ghettos to follow the Lane with gold all the way to our ghetto. Our ghetto owner is now trying to get the people and tools ready again. Definitely we will get there and hopefully we will get a lot.

This story shows the *social–political–geological* three-dimensional orientation of his mining team. To wit, a social strategy (the strike) is interrupted by a geological urgent situation (the Lane) and the anticipation of a social conflict because of the Lane. The case also touches upon a perception of gold that is typical for many people in various West African countries. I will elaborate shortly on this because it enables me to discuss some of the *cosmological* three-dimensional orientations of artisanal gold miners in Kejetia regarding gold, land and gold mining – and to compare that with literature on miners' cosmological perceptions in various West African countries.

The miners' remark about the 'vanished money' corresponds with a perception that artisanal gold miners in Burkina Faso have, namely: that money acquired through gold is 'bitter money'.[50] Following this perception, 'bitter money' cannot be the basis for lasting prosperity. The reason for this is that gold belongs to ancestors and land spirits – a perception that is also common among the Malinke (who live in multiple West African countries).[51] According to the idea of 'bitter money', ancestors will punish the seller of the gold by 'following' the money flow that originated from the gold. That is why money acquired through gold, according to the idea of 'bitter money', should be spent on luxury goods rather than on primary needs. Werthmann explains that people in Burkina Faso do continue to be involved in artisanal gold mining for several reasons: poverty (Jansen nuances this),[52,53] the belief that ancestors and spirits can be satisfied with sacrifices and because younger generations attach less value to the idea of 'bitter money'.[54] Artisanal gold miners in Kejetia also relate spirits and ancestors to gold and the underground – the tindana has a role in intervening in these relations, which is illustrated later in this section. The miners spend the money acquired through gold in various ways. Home building, marriage and school were common answers to the questions about why people were involved in artisanal gold mining – which differs from the idea that money acquired from gold is 'bitter money' and should not be spend on primary needs. However, many artisanal gold miners also spend lots of the acquired money on alcohol and marijuana. Artisanal gold miners explain that they do this in order 'to get the strength to do the hard work and go under the ground again'. They often hope to, one day, find a lot of big gold nuggets that would enable them to go back home rich (and not have to come back to Kejetia again).

It is remarkable that some artisanal gold miners in Kejetia, who do relate gold to spirits and ancestors, also refer to the extraction of gold as *cultivation*. This seems to move away from the concept of *hunting* for gold, which the above-mentioned Malinke relate to the idea that gold belongs to spirits and ancestors. To wit, gold mining requires the skill to both work in the bush and to please the spirits while stealing the gold, as is the case with hunting activities.[55] This approach to gold mining implies that gold has a wild nature. However, perceiving gold mining as cultivation seems to imply a much tamer nature, making gold mining (from a cosmological perspective) less dangerous and more like harvesting.

The relation between microgovernance and cosmological perceptions is noticeable in (among other things) the taboo in and around Kejetia on mining on Fridays. Some artisanal gold miners in Kejetia mention this taboo: on Friday, the land has to rest and the ancestors (the owners of the gold) should be left alone. However, some artisanal gold miners choose to go under the ground to mine on Friday anyway – also, not all artisanal gold miners in Kejetia mention ancestors and spirits in relation to gold, land and their gold mining activities.

An anecdote about a collapsed ghetto illustrates the cosmological perceptions of gold and land in Kejetia and their influence on governance: in November 2014, a ghetto in Obuasi (Kejetia's neighbouring artisanal gold mining community) collapsed when artisanal miners were under the ground. Some artisanal gold miners claimed that this happened because the miners in the collapsed ghetto went mining on a Friday. According to them, the miners in the collapsed ghetto should not have disturbed the land spirits. The tindana (who knows much about the land, land spirits and ancestors) found out that the ancestors and spirits of the land had not been satisfied and he sacrificed a cow after the collapse of the ghetto to satisfy the ancestors and land spirits again. The tindana works together with the chief. When the tindana believes the underground is safe to work again, the chief allows artisanal miners to continue their work.

As people in Kejetia come from very different places (within Ghana and from neighbouring countries), perceptions and beliefs regarding land and gold are diverse. However, during the research, it became clear that for all artisanal gold miners in Kejetia, the possibilities of acquiring gold are embedded in a web of social and cosmological relations and obligations that go beyond actual excavation. This makes gold a *substance* for artisanal gold miners rather than a *resource*.

An artisanal miner gives another example of geological structures that create social tension:

> In our ghetto we have gutter stone and ceiling stone (Table 1). When we get ceiling stone, we have to be fast! Because it contains a lot of gold and it moves towards the ghettos of other miners; the risk is that other mining teams might find it before we do, because the gold is closer to the surface a few meters from our ghetto (where the other teams ghetto is).

In this case, the direction of the gold line (going up) increases the chance that other mining teams will reach the gold line before this miner's team can cultivate the gold. This creates urgency within the mining team and a possible underground conflict with the other mining team in which the chief would have to mediate.

The subterranean water level is also an issue for artisanal gold miners in Kejetia. A shop owner and sponsor states that the water in ghettos limits the work of artisanal gold miners:

> At the moment, there are only a few people in Kejetia because there is no work now. There is water in the ghettos, so they first have to pull out the water. They have been pulling the water out of the ghettos for three days now, I think after another four days they are done. When the water finishes, a lot of people will come. It will be busy in Kejetia! A lot more busy than now.

When a mining team finally reaches more profitable gold ore, the team has to assess what kind of gold ore they have reached. The subterranean geology determines what strategy the miners prefer to use and what kind of mining expertise is needed. For example, a mining team might reach a Bulk stone. The ghetto owner generally prefers the *chisel man* to chisel away the useful ore because this saves the cost of the dynamite (Table 1). However, a hammer and chisel work for Package stone, not for Bulk stone so it is up to the *blast man* and *drill man* to separate the rock (waste) from the stone with dynamite. The *local boys* have the job of carrying the waste up to the surface. Ghetto owners in some cases decide to leave the Bulk stone hanging under the ground for a while. This can be for several reasons. First of all, he does not want the stone to get mixed up with the waste. Secondly, the team wants to take the stone down together, with everybody there, in order to make sure that the stone is fairly divided. Finally, when a mining team waits to take up a large amount of stone, they can keep a bigger total portion for themselves. To wit, every time an artisanal mining team takes up stones, they have to give something to the chief, tindana and the sponsor – furthermore, if a team can process many stones at the same time and thus get more gold at the same time, they can use a piece to sell to (and satisfy) their sponsor and also use a piece to hide from their sponsor and sell to someone they do not owe, for a higher price. Some mining teams have a *security man* to prevent the artisanal miners that work under the ground from taking little pieces of stone or to protect Lanes from other mining teams.

Above the ground: processing the gold ore

When the Bulk stone has been taken down, all team members get a share of the stones (Table 1). They process their stones together, separately or in groups – this is different for every team. The concession and ghetto owner get a large portion of the stones. Then, it is time to process the stones in order to acquire the gold. The main activities of gold ore processing are crushing and grinding the stones, washing the stone dust and using mercury to capture the gold from the last mix of water, stone dust and gold dust. The above-ground gold ore processing activities, in which both women and men participate, are for a large part based on the substance features of gold: weight, smelt temperature and the ability to attach to mercury. The substance features of gold, and the gold ore processing techniques required for those features, determine a large part of the above-ground social organisation of the artisanal gold mining community in Kejetia.

Conclusion

Using in-depth ethnographic analysis of Kejetia's artisanal gold mining community in Tongo (Northern Ghana), this article argues that governance is based on artisanal gold miners' three-dimensional orientation towards both 'above-ground' social–political relations and regulations (a geometrically two-dimensional orientation) and the 'subterranean' geological sphere (which literally adds depth and a third geometric dimension). A substance relation to gold also means that its properties and geological context strongly influence miners' microgovernance, as do miners' cosmological perceptions of gold's relationship to the above-ground and subterranean spheres. Hence, gold acquisition is embedded in a web of geological, social and cosmological relations beyond actual excavation. Artisanal mining policy and research should recognise the impact of this three-dimensional orientation and substance approach.

The complex multidimensional regulations, in which artisanal gold miners have to engage, make it important for artisanal gold miners to position themselves well, socially and politically, above-ground. Above-ground social–political (f)actors, on which Kejetia's miners orientate themselves, are mainly the chief, the tindana, the Minerals Commission of the Ghanaian Government, concession leaders, competing mining teams and the issue of 'being autochthonous' or 'being a stranger'. Orientation on these (f)actors enables miners to (strategically) undertake their (subterranean) mining activities. Claims to land and gold in Kejetia are legitimised through various mechanisms. Artisanal miners who are autochthonous (from Tongo) generally legitimise their claims in terms of 'who was first'. They often have a strong relation to the chief in Kejetia, which is beneficial in subterranean conflicts. Artisanal gold miners, especially Ghanaian miners that are perceived as strangers in Kejetia, also use the legal framework of the national government (sometimes through arrangements with concession leaders) to claim land and gold in Kejetia. Above-ground orientation is always combined with subterranean orientation.

Subterranean geological factors and substance features of gold are as well part of Kejetia's artisanal gold miners' orientation. The organisation and strategies of subterranean mining tasks and teams are very much based on subterranean geological structures. The direction of subterranean gold veins are in particular strongly related to social tensions and conflicts: Is a gold vein moving to the direction of another ghetto?, Is another mining team coming in the direction of your ghetto by following a gold vein?, Is your team going to encounter another mining team under the ground and what will then happen?[56] Above-ground gold ore processing activities also make up a large part of the social organisation in Kejetia and are based for a large part on substance features of gold: weight, smelt temperature and the ability to attach to mercury.

Artisanal gold miners' cosmological perceptions of gold and land also contribute to microgovernance in Kejetia. Artisanal gold miners often view gold as belonging to ancestors or land spirits. The possibilities for Kejetia's artisanal gold miners to acquire gold are embedded in a web of social and cosmological obligations and relations – which are typical for social–political relations in Northern Ghana – beyond actual excavation. This means that to understand organisation and microgovernance in artisanal gold mining communities, we should research this web of social and cosmological relations and approach gold as a substance and not as a resource. This approach would enable understanding of social, cultural, cosmological and geological perceptions as well as economical perceptions of gold.[57]

The three-dimensional orientation of Kejetia's artisanal gold miners differs from the orientation of most policy-makers. Policy-makers tend to orientate themselves socio-politically and above ground. In order to deal adequately with artisanal miners, policy measures and studies that consider artisanal mining should comprehend the impact of the three-dimensional orientation and substance approach of artisanal miners on the dynamics of their mining behaviour. Policy-makers and researchers should therefore be aware of the pitfalls of literally overlooking subterranean contributions to above-ground artisanal mining micro governance.

Disclosure statement

No potential conflict of interest was reported by the author.

Funding

This work was supported by Leids Universitair Fonds (LUF) – LISF, [grant number 14106].

Acknowledgements

I sincerely thank the anonymous reviewers of *Third World Thematics* for their feedback and contribution to the article. I am also grateful for the great supervision of Jan Jansen (Leiden University) throughout all phases of the research and for his insights and comments to improve this article. Furthermore, I thank Chris Huggins for his contributions to the article. Finally, I want to express my sincere gratitude to Tongdan Abdallah Kwadanaab, the Tallensi community in Tongo and the artisanal gold miners in Kejetia for enabling my research in Tongo and Kejetia.

Notes

1. Hilson, 'Foreword', x.
2. Hilson, 'Structural Adjustment in Ghana', 54, 67.
3. See note 1 above.
4. Lund, '"Bawku is Still Volatile"', 587; Geschiere, *The Perils of Belonging*, 16–19; Bavinck et al., 'Introduction', 6; Wilson and Tisdell, *Conflicts, Resources and Environment*, 1.
5. Grätz, 'Gold Trading Networks', 152.
6. Banchirigah and Hilson, 'De-Agrarianization, Re-Agrarianization and Local Economic', 157.
7. Hilson, 'Foreword', xi.
8. Ostrom, 'A General Framework', 419.
9. Ballard and Banks (Godoy), 'Resource War', 307; Panella, 'Gold Mining in West Africa', 2–3; Luning and Pijpers, 'Co-habitation in 3D', 1.
10. Boons, 'Organizing Within Dynamic Ecosystems', 284.
11. See note 10 above; See note 8 above.
12. Ostrom furthermore presents a framework for the social–ecological system. She connects (S) social and cultural settings, (ECO) the related ecosystems, (U) users, (RU) the resource units, (RS) the resource system and finally (GS) the governance system. She presents as well 10 variables that influence the way in which people locally self-organise and structure themselves around resources: (1) size of resource system, (2) productivity of system, (3) predictability of system dynamics, (4) resource unit mobility, (5) number of users, (6) leadership, (7) norms/social capital, (8) knowledge of the SES, (9) importance of resource to users and (10) collective choice rules. I do not use the terms in this framework. However, the notion that all elements in a social–ecological system are constantly influencing each other is important in this article.
13. See note 10 above.
14. Barnes et al., 'Contribution of Anthropology', 541.
15. Ibid., 543.
16. Bleischwits et al., 'Coltan from Central Africa', 5.
17. Ballard and Banks, 'Resource War', 307.
18. Gewald, 'Gold as a Geological Item', 15.
19. Panella, 'Gold Mining in West Africa', 2–3.
20. Pijpers, 'Busting Diamonds, Booming Gold', 83.
21. Luning and Pijpers, 'Co-habitation in 3D', 1, 22.
22. Puijenbroek et al., *A Golden Future in Ituri?* 19.
23. Aryee, 'Small-Scale Mining Ghana', 372.
24. See note 15 above.
25. Banchirigah and Hilson, 'De-Agrarianization, Re-Agrarianization and Local Economic Development', 1.
26. Ballard and Banks, 'Resource War', 297.
27. These stakeholders, and the developments of these stakeholders, turn out to be very much intertwined.
28. Orlove and Caton, 'Water Sustainability', 401, 404, 408.

29. In the period of January–March 2015, multi-sited anthropological ethnographic field research was conducted in Tongo and Kejetia. While living with both the artisanal gold miners in Kejetia and with the political elite of the Tallensi in the Tongo centre, participant observation, semi-structured interviews and participatory mapping methods were the main research methods. Semi-structured interviews were done with artisanal gold miners (with various backgrounds and functions in Kejetia, both men and women), the paramount chief of Tongo, the representative chief of Kejetia, the representative chief of Obuasi, sponsors and buyers in Kejetia, the HR and PR manager of Shaanxi Mining Company (GH) Ltd. and with the Upper East Director of the Environmental Protection Agency in Ghana. There was no translator used.
30. Kejetia is named after the large market in Kumasi.
31. Long et al., 'Water Values in a Ghanaian', 209.
32. Ibid., 201.
33. Ibid.
34. Pijpers (Werthmann), 'Busting Diamonds, Booming Gold', 85–88.
35. This happened as a result of an intervention to solve a dispute in the village GbokGbani. The Tongo paramount chief separated the village and made Gbok and Gbani. Gbok and Gbani are now 2 of the 36 villages that are part of Tongo and the Tallensi district.
36. The governmental agencies Minerals Commission and Environmental Protection Agency work (together) to establish a hold on artisanal and small-scale mining through laws such as the *Minerals and Mining Act* (Act 703, introduced in 2006) ('As of July 20, 2015, the Ghana Extractive Industries Transparency Initiative listed on its website;''Minerals and Mining Act 703' Accessed July 2015. http://www.sdsg.org/wp-content/uploads/2011/06/Ghana-Minerals-Act-2006.pdf: point 81–99). This law emphasises the mineral sovereignty of the Ghanaian state and obliges artisanal miners to register for a small-scale mining concession.
37. Hilson, 'Structural Adjustment in Ghana', 60.
38. Hilson and Potter, 'Structural Adjustment and Subsistence Industry', 108.
39. Luning and Pijpers (Newmont), 'Co-habitation in 3D', 3.
40. Hilson, 'Structural Adjustment in Ghana', 53, 57, 59–60.
41. In 1983, the Ghanaian Government started the national ERP under the guidance of the IMF and World Bank. This happened after the Ghanaian economy deteriorated in the 1970s and early 1980s and, by 1983, was in a state of crisis. Since 1983, the Ghanaian Government developed policies to establish a more attractive investment climate for foreign mineral exploration and extraction companies. The ERP was meant to privatise the mineral sector of Ghana and make it more accessible and attractive for foreign investors. The arrival of Shaanxi in Tongo was also initiated among others by the national government.
42. The agreement and collaboration between Shaanxi and the representative chief in Obuasi (who is both a local authority and living in Kejetia with his artisanal mining brothers) are facilitated by the Ghanaian national government. This arrangement is an example of the fact that the different kind of stakeholders (identified by Ballard and Banks, as described above) that are involved in mining communities are very much overlapping and intertwined. Also, Shaanxi mines on a small-scale mining concession, while performing large-scale and industrial mining activities.
43. Geschiere, *The Perils of Belonging*, 16–19.
44. Luning and Pijpers (Newmont), 'Co-habitation in 3D', 3.
45. The actual chief and tindana live in Gbani. The mining area Kejetia is part of the village Gbani (in the Tongo region). In Northern Ghana, chiefs are typically in charge of above-ground political matters, whereas tindana typically manage the relations that people have with underground land spirits and ancestors.
46. Aryee, 'Small-Scale Mining Ghana', 370.
47. In Kejetia, I met no women who were concession leaders.
48. Concession leaders might suggest strategic temporary mining breaks, to the chief or tindana, for specific mining areas at moments this would suit them.
49. This is what many artisanal gold miners in Kejetia claim, I never saw actual contracts.
50. Werthmann, 'Cowries, Gold and "Bitter Money"', 105.
51. Keita, 'Les Sacrifices pour l'Orpaillage', 127.

52. Jansen, 'What Gold Mining Means', 97–98.
53. Jansen explains that migrations to artisanal gold mining sites in West African countries (like Mali and Guinea) are not an economic activity per se. In several West African countries, every dry season, thousands of people migrate to gold mining sites. These migrations to gold mining sites are not an economic activity as they have a social function as well. The migration enables people to establish a safety network for current or future periods of scarcity. Relations around the gold mines are organised in terms of hosts and guests – people can be a guest one year, and a host the next year, as the locations of gold mining sites might change over time.
54. Werthmann, 'Cowries, Gold and "Bitter Money"', 118–119.
55. Jansen, 'What Gold Mining Means', 99–100.
56. It appears here that the strategy of artisanal miners to determine the location for their ghetto – the strategy of reading the locations of other ghettos (the rows of ghettos) – will inevitably lead to mining teams meeting under the ground. Some people living in Kejetia claimed that artisanal gold miners sometimes use metal detectors to determine the location of their ghetto, although I did not observe this. It would be interesting to know whether or not the strategy of not following the rows of ghettos leads to fewer (or later) meetings under the ground.
57. See note 26 above.

Bibliography

Aryee, Benjamin. "Small-Scale Mining in Ghana as a Sustainable Development Activity: Its Development and a Review of the Contemporary Issues and Challenges." In *The Socio-Economic Impacts of Artisanal and Small Scale Mining in Developing Countries*, edited by Gavin Hilson, 360–398. Lisse: A.A. Balkema Publishers, 2003.

Ballard, C., and G. Banks. "Resource Wars: The Anthropology of Mining." *Annual Review of Anthropology* 32 (2003): 287–313.

Banchirigah, S. M., and G. Hilson. "De-Agrarianization, Re-Agrarianization and Local Economic Development: Re-Orientating Livelihoods in African Artisanal Mining Communities." *Policy Sciences* 43, no. 2 (2010): 157–180.

Barnes, J., D. Michael, M. Lahsen, A. Mathews, P. McElwee, R. McIntosh, F. Moore, et al. "Contribution of Anthropology to the Study of Climate Change." *Nature Climate Change* 3 (2013): 541–544.

Bavinck, Maarten, Lorenzo Pellegrini, and Erik Mostert. "Introduction." In *Conflicts Over Natural Resources in the Global South – Conceptual Approaches*, edited by Maarten Bavinck, Lorenzo Pellegrini and Erik Mostert, 1–11. Amsterdam: CRC Press, Taylor & Francis Group, 2014.

Bleischwitz, R., M. Dittrich, and C. Pierdicca. "Coltan from Central Africa, International Trade and Implications for Any Certification." *Resources Policy* 37, no. 1 (2012): 19–29.

Boons, F. "Organizing Within Dynamic Ecosystems: Conceptualizing Socio-Ecological Mechanisms." *Organization & Environment* 26, no. 3 (2013): 281–297.

Geschiere, P. L. *The Perils of Belonging – Autochthony, Citizenship, and Exclusion in Africa & Europe*. Chicago, IL: The University of Chicago Press, 2009.

Gewald, H. "Gold as a Geological Item." In *Worlds of Debts: Interdisciplinary Perspectives on Gold Mining in West Africa*, edited by Cristiana Panella, 15–24. Amsterdam: Rozenberg Publishers, 2010.

Grätz, T. "Gold Trading Networks and the Creation of Trust: A Case Study from Northern Benin." *Africa: Journal of the International African Institute* 74, no. 2 (2004): 146–172.

Hilson, G. M., and C. Potter. "Structural Adjustment and Subsistence Industry: Artisanal Gold Mining in Ghana." *Development and Change* 36, no. 1 (2005): 103–131.

Hilson, G. M. "Structural Adjustment in Ghana: Assessing the Impacts of Mining-Sector Reform." *Africa Today* 51, no. 2 (2004): 53–77.

Hilson, G. M. "Foreword." In *The Socio-Economic Impacts of Artisanal and Small Scale Mining in Developing Countries*, edited by Gavin Hilson, x–xii. Lisse: A.A. Balkema Publishers, 2003.

Jansen, Jan. "What Gold Mining Means for the Malinke, and How it was Misunderstood by the French Colonial Administration." In *Worlds of Debts: Interdisciplinary Perspectives on Gold Mining in West Africa*, edited by Cristiana Panella, 95–110. Amsterdam: Rozenberg Publishers, 2010.

Keita, Mahamadou. "Les Sacrifices pour l'Orpaillage Artisanal dans le Manding." In *Worlds of Debts: Interdisciplinary Perspectives on Gold Mining in West Africa*, edited by Cristiana Panella, 127–136. Amsterdam: Rozenberg Publishers, 2010.

Long, R., E. Renne, T. Robins, M. Wilson, K. Pelig-Ba, M. Rajaee, A. Yee, et al. "Water Values in a Ghanaian Small-Scale Gold Mining Community." *Human Organization* 72, no. 3 (2013): 199–210.

Lund, C. "'Bawku is Still Volatile': Ethno-political Conflict and State Recognition in Northern Ghana." *The Journal of Modern African Studies* 41, no. 4 (2003): 587–610.

Luning, Sabine, and Robert Pijpers. "Cohabitation in 3D: Governing Gold Concessions in Ghana." *Africa: Journal of the International African Institute*, forthcoming.

Orlove, B., and S. C. Caton. "Water Sustainability: Anthropological Approaches and Prospects." *Annual Review of Anthropology* 39 (2010): 401–415.

Ostrom, E. "A General Framework for Analyzing Sustainability of Social-Ecological Systems." *Science* 325 (2009): 419–422.

Panella, Cristiana. "Gold Mining in West Africa – Worlds of Debts and Sites of Co-habitation." In *Worlds of Debts: Interdisciplinary Perspectives on Gold Mining in West Africa*, edited by Cristiana Panella, 1–14. Amsterdam: Rozenberg Publishers, 2010.

Pijpers, Robert. "Busting Diamonds, Booming Gold: Crisis, Gender, and Money in the Sierra Leone Mining Sector." In *Worlds of Debts: Interdisciplinary Perspectives on Gold Mining in West Africa*, edited by Cristiana Panella, 79–94. Amsterdam: Rozenberg Publishers, 2010.

van Puijenbroek, Joost, Eric Mongo Malolo and Jogien Bakker. *A Golden Future in Ituri? Which Perspective for Gold Exploitation in Ituri, DR Congo?*: IKV PAX Christi & Réseau Haki Na Amani, Utrecht, 2012.

Werthmann, K. "Cowries, Gold and 'Bitter Money': Gold-Mining and Notions of Ill-Gotten Wealth in Burkina Faso." *Paideuma* 49 (2003): 105–124.

Wilson, C., and C. Tisdell. *Conflicts Over Natural Resources and the Environment Economics and Security*. Brisbane: The University of Queensland, 2003.

Index

Note: Page numbers in *italic* refer to figures
Page numbers in **bold** refer to tables
Page numbers followed by 'n' refer to notes

For Product Safety Concerns and Information please contact our EU
representative GPSR@taylorandfrancis.com
Taylor & Francis Verlag GmbH, Kaufingerstraße 24, 80331 München, Germany

www.ingramcontent.com/pod-product-compliance
Lightning Source LLC
LaVergne TN
LVHW080656011026
849160LV00014B/1455

* 9 7 8 1 0 3 2 0 9 0 4 5 0 *